New Insights on Principal Component Analysis

Edited by Fausto Pedro García Márquez, Mayorkinos Papaelias and René-Vinicio Sánchez Loja

Published in London, United Kingdom

New Insights on Principal Component Analysis
http://dx.doi.org/10.5772/intechopen.111238
Edited by Fausto Pedro García Márquez, Mayorkinos Papaelias and René-Vinicio Sánchez Loja

Contributors
Ximing Chen, Anthony O. Smith and Anand Rangarajan, Wataru Souma, Atallah Ahmed, Touati Hayat, Saad Mohammed Abdelmoudjib, Amrani Amel, Berrabah Ameur, Cherifi Selma, Allali Taleb and Benkhaled Hadj, Oswaldo Eduardo Ramos Ramos and Leonardo Guzmán Alegría, Esteban Vegas, Lluís Serra, Ferran Reverter and Josep Maria Oller, Miantsia Fokam Olivier, Felix Meutchieye and Evaristus Tsi Angwafo

First published in London, United Kingdom, 2024 by IntechOpen
IntechOpen is the global imprint of INTECHOPEN LIMITED, registered in England and Wales, registration number: 11086078, 5 Princes Gate Court, London, SW7 2QJ, United Kingdom

British Library Cataloguing-in-Publication Data
A catalogue record for this book is available from the British Library

Additional hard and PDF copies can be obtained from orders@intechopen.com

New Insights on Principal Component Analysis
Edited by Fausto Pedro García Márquez, Mayorkinos Papaelias and René-Vinicio Sánchez Loja
p. cm.
Print ISBN 978-0-85466-266-1
Online ISBN 978-0-85466-265-4
eBook (PDF) ISBN 978-0-85466-267-8

For EU product safety concerns:
IN TECH d.o.o., Prolaz Marije Krucifikse Kozulić 3, 51000 Rijeka, Croatia,
info@intechopen.com or visit our website at intechopen.com.

Meet the editors

Fausto Pedro García Márquez is a full professor at the University of Castilla-La Mancha (UCLM), Spain. He is also an honorary senior research fellow at Birmingham University, UK; a lecturer at the Postgraduate European Institute, Spain; a research fellow at INTI International University & Colleges, Malaysia; and a senior manager at Accenture. He obtained his European Ph.D. with maximum distinction. He has been distinguished with numerous prizes, including Runner Prize (2023), Nominate Prize (2022), Gran Maestre (2022), Grand Prize (2021), Runner Prize (2020), Advancement Prize (2018), and Runner (2015), Advancement (2013), and Silver Prizes (2012) by the International Society of Management Science and Engineering Management (ICMSEM). He is also the recipient of the 2017 First International Business Ideas Competition. He has published more than 257 journal papers and authored/edited 53 books and more than 100 book chapters. He also has six patents to his credit. He is the editor of five international journals and a committee member for more than seventy international conferences. He has been the principal investigator for four European projects, eight national projects, and more than 150 projects for universities and companies. His main interests are artificial intelligence, maintenance, management, renewable energy, transport, advanced analytics, and data science. He is an expert for the AI4People project by Atomium – the European Institute for Science, Media and Democracy (EISMD) and the European Science Foundation (ESF). He is also the director of the Igenium research group, a senior member of the Institute of Electrical and Electronics Engineers (IEEE), an honorary member of the Research Council of the Indian Institute of Finance, and a committee chair of the International Society for Management Science and Engineering Management (ISMSEM).

Professor Mayorkinos Papaelias has a Ph.D. in Metallurgy from the University of Birmingham, England. He leads research in non-destructive testing and structural health condition monitoring at the School of Metallurgy and Materials and the Birmingham Railway Centre for Research and Education. He conducts research in structural health condition monitoring of wind turbine towers and advanced condition monitoring of wind turbine gearboxes and rotating machinery. He served as a technical consultant to TWI Ltd,, ENGITEC Systems International, Innovative Technology and Science Ltd, and the Institute of Welding and Quality. He is the editor of four books on fault detection and condition monitoring and has contributed chapters to books on fault detection and rail inspection. Dr. Papaelias is chairman of the Education Committee of the International Society for Condition Monitoring of the British Institute of Non-Destructive Testing.

René-Vinicio Sánchez is a full professor at the Universidad Politécnica Salesiana (UPS), Ecuador. He obtained his master's degree and Ph.D. in Industrial Technologies Research from the Universidad Nacional de Educación a Distancia (UNED), Spain. He also has a Ph.D. in Engineering from the Universidad Pontificia Bolivariana, Colombia. Since 2004, he has been a professor at UPS, mainly in areas related to sequential process automation. In 2014, he founded the Industrial Technologies Research and Development Group (GIDTEC) and completed a guest Ph.D. abroad at Chongqing University of Technology and Business, China. He has been a senior member of the Institute of Electrical and Electronics Engineers (IEEE) since 2022. He has extensive experience in conference organization, technology project implementation, and research project management and execution. Currently, he has more than sixty publications to his credit. He is also a reviewer for several high-impact journals. His current interests include project management, condition-based maintenance, engineering education, and Industry 4.0, especially for small and medium-sized enterprises (SMEs).

Contents

Preface

In contemporary industries, the utilization of extensive datasets has become customary and is on the rise. This surge is attributed to the integration of new sensors, more intricate systems, and the imperative to enhance system reliability, availability, and safety. The Internet of Things (IoT) serves as another significant data source, furnishing a vast array of varied data.

Principal Component Analysis (PCA) is experiencing heightened adoption and advancements, either independently or in conjunction with other methodologies, for analyzing these datasets. PCA primarily functions by altering the dataset through a linear transformation, reducing the coordinate system. This transformation generates a new set of coordinates termed "principal components," derived based on variance, with the first principal component representing the highest variance.

The fundamental aim of PCA is to condense the size of a dataset into a smaller transformed space governed by the eigenvectors associated with the original dataset's covariances. These eigenvectors are ranked based on their maximum variability, and thus, are termed principal components. Essentially, this method reshapes the initial dataset into a new p-dimensional set of Cartesian coordinates, a projection onto the principal component vector, with direction guided by the P matrix, where "a" denotes the largest eigenvalue and its columns represent the retained eigenvectors.

PCA can also be linked to other algorithms, such as factor analysis, non-negative matrix factorization, correspondence analysis, and K-means clustering, among others. Moreover, PCA has evolved, giving rise to alternative algorithms that address certain limitations, like Sparse Principal Component Analysis, Robust Principal Component Analysis, or Nonlinear Principal Component Analysis. Furthermore, PCA has demonstrated efficacy when combined with other algorithms, as previously mentioned.

This book features contributions from various authors, consolidating analytical principles with business applications. It explores the relationship between core disciplines like technology, engineering, and organizational abilities, showcasing PCA's applications. It also encompasses diverse specialties like finance, risk analysis, marketing, and economics. The book elucidates practical case studies across multiple industries employing PCA, ranging from straightforward to highly complex problem-solving scenarios, encompassing static, dynamic, and large-scale problems.

Fausto Pedro García Márquez
Ingenium Research Group,
University of Castilla-La Mancha,
Ciudad Real, Spain

Mayorkinos Papaelias
University of Birmingham,
Birmingham, United Kingdom

René-Vinicio Sánchez Loja
Research and Development Group in Industrial Technologies,
Salesian Polytechnic University,
Cuenca, Ecuador

Chapter 1

Acoustic Emission Signal Processing Method and Modern Modeling Technology

Ximing Chen

Abstract

Using acoustic emission signal as the detection medium for particle characteristic parameters has the advantages of real-time, non-destructive, safe, and non-invasive flow field. In order to extract the rich information contained in the acoustic emission signal and establish the quantitative relationship between the acoustic emission signal and the particle characteristic parameters, it is necessary to carry out a series of mathematical processes on the acoustic emission signal in order to extract valuable features from it, and then take these features as the model variables and, through modern modeling methods, establish the quantitative relationship between the acoustic emission signal mode characteristics and the particle characteristic parameters. This chapter first introduces the application status and research progress of acoustic emission technology in chemical processes, then introduces the processing methods of acoustic emission signals, and finally focuses on the basic principles of wavelet (packet) analysis, the types of wavelet (packet) functions, the Mallat algorithm, signal wavelet (packet) noise reduction, and other basic theories, as well as the research progress of particle detection based on modern modeling technology of acoustic emission signals.

Keywords: Fourier transformation, wavelet, wavelet packet, chemical production, acoustic emission signal

1. Introduction

Fourier analysis can only provide the frequency of the signal in the whole time domain, but it cannot provide the frequency information of the signal in a certain time period. Short-time Fourier transform divides the whole time domain into some small, equal time intervals, and then, Fourier analysis is used in each time period. It contains time and frequency information to a certain extent, but because the time interval cannot be adjusted, it is difficult to detect the short duration, the time when a pulse signal with high frequency occurs [1].

Wavelet are mathematical tools for analyzing time series or images [2].

The concept of wavelet transform was proposed first by J. Morlet, a French engineer engaged in petroleum signal processing, in 1974, but was not recognized by mathematicians at that time. In 1986, the famous mathematician Y. Meyer accidentally constructed a true wavelet basis and collaborated with S. Mallat to establish a unified method for constructing wavelet bases with multi-scale analysis [3, 4]. After that, wavelet analysis began to flourish. Among them, the Belgian female mathematician I. Daubechies' "Ten Lessons on Wavelets" played an important role in promoting the popularization of wavelets [5]. Compared with the Fourier transform and window Fourier transform (Gabor transform), it is a local transform of time and frequency that can effectively extract information from signals. Through operational functions such as scaling and translation, it performs multi-scale analysis on functions or signals, solving many difficult problems that cannot be solved by the Fourier transform. Therefore, wavelet transform is known as the "mathematical microscope." It is a milestone in the development history of harmonic analysis.

This section focuses on the basic theory of wavelet analysis and the method of signal time-frequency multiscale decomposition.

2. The basic principles of wavelet analysis $L^2(R)$ space

The function space mainly discussed in wavelet analysis is a real function space composed of square-integrable function $L^2(R)$. That is, $f(t) \in L^2(R) \Leftrightarrow \int_R |f(t)|^2 dt < +\infty$, it is an infinite dimensional vector space. One of the main problems in wavelet analysis research is how to represent functions in space $L^2(R)$ using the dyadic contraction and translation of a basic wavelet function, that is: $f(t) = \sum_{n \in Z} b_n \phi_n(t)$.

2.1 Continue wavelet transform

Expanding the function $f(t)$ in any space $L^2(R)$ on a wavelet basis is called a continuous wavelet transform of the function $f(t.)$

Definition 1.1

Assume that the function $f(t) \in L^2(R), \psi(t) \in L^2(R)$, and $\psi(t)$ satisfies the permissibility condition:

$$C_\psi = \int_{-\infty}^{+\infty} \frac{|\hat{\psi}(\omega)|^2}{|\omega|} d\omega < +\infty \tag{1}$$

where $\hat{\psi}(\omega)$ is the Fourier transform of the function $\psi(t)$, that is, $\hat{\psi}(\omega) = \int_{-\infty}^{+\infty} \psi(t) e^{-i2\pi\omega t} dt$. The continuous wavelet transform of $f(t)$ is defined as:

$$\boldsymbol{WT}_x(a,b) = <f, \psi_{a,b}> = \frac{1}{\sqrt{a}} \int_{-\infty}^{+\infty} f(t) \psi^* \left(\frac{t-b}{a}\right) dt, \quad a \neq 0 \tag{2}$$

In the formula (2), "*" represents taking conjugation to a complex number, where

$$\psi_{a,b} = \frac{1}{\sqrt{a}} \psi \left(\frac{t-b}{a}\right) a \neq 0 \tag{3}$$

For the wavelet generating function $\psi(t)$, it can be seen from condition (1), $\hat{\psi}(0) = 0$ is necessarily (if $\hat{\psi}(0) \neq 0$, then $C_\psi = \infty$, admissibility condition cannot be satisfied), so it follows that:

$$\hat{\psi}(0) = \int_{-\infty}^{+\infty} \psi(t)dt = 0 \tag{4}$$

That is, the algebraic sum of $\psi(t)$ and the area enclosed by the entire horizontal axis is zero, having bandpass properties. Its graph appears as an alternating positive and negative oscillating waveform on the horizontal axis, hence it is called "wavelet."

Any linear transformation used for signal reconstruction should meet the requirement of complete reconstruction, and the same applies to wavelet transform. The reconstruction formula for wavelet transforms that meet the allowable conditions is:

$$f(t) = C_\psi^{-1} \int_{-\infty}^{+\infty} \int_{-\infty}^{+\infty} \boldsymbol{WT}_x(a,b)\boldsymbol{\psi}_{a,b}(t)\frac{da}{|a|^2}db \tag{5}$$

Continuous wavelets have the following properties [6]:

1. **Linearity:** the wavelet transform of a multi-component signal is equal to the sum of the wavelet transforms of each component, which can be expressed as follows by the formula:

$$f(t) = \sum_{i=1}^{n} f_i(t)$$

$$f(t) \leftrightarrow WT, f_i(t) \leftrightarrow WT_i$$

then: $WT = \sum_{i=1}^{n} WT_i$

2. translation invariance (timeshift covariation):

$$iff(t) \leftrightarrow WT_x(a,b), thenf(t-\tau_0) \leftrightarrow WT_x(a, b-\tau_0)$$

3. Time scale theorem (dilation covariance):

$$\text{if} f(t) \longleftrightarrow WT_x(a,b), \text{ then} f(ct) \longleftrightarrow \frac{1}{\sqrt{C}} WT_x(ca,b), c > 0$$

4. Self-similarity:

CWT (Continuous wavelet transform) transforms one-dimensional signals into two-dimensional space $f(t) \leftrightarrow WT_f$. Therefore, there is redundant information in wavelet transform called redundancy. The continuous wavelet transforms corresponding to different scale parameters a and different translation parameters b are self-similarity. There is redundancy in information representation in continuous wavelet transform.

That is to say, the inverse transformation of wavelet transform is not unique. $\psi_{a,b} = \frac{1}{\sqrt{a}}\psi\left(\frac{t-b}{a}\right)$ is a family of super complete basis functions, they are linearly correlated. The measure of redundancy is called the regenerative kernel $K(a_0, b_0, a, b)$:

$$K(a_0, b_0, a, b) = \frac{1}{C_\psi}\int_R \psi_{a,b}(t)\psi^*_{a_0,b_0}(t)dt = \frac{1}{C_\psi}\int_R \frac{1}{\sqrt{a}}\psi^*_{a,b}\left(\frac{t-b}{a}\right)\psi^*_{a_0,b_0}\left(\frac{t-b_0}{a_0}\right)dt$$
$$= \frac{1}{C_\psi}\langle \psi_{a,b}(t), \psi^*_{a_0,b_0}(t)\rangle \tag{6}$$

$C_\psi < \infty$ is needed in order to achieve the complete refactoring. This is also the permissibility condition mentioned in Eq. (1), also known as the complete reconstruction condition.

As the implementation of signal reconstruction is numerically stable, in addition to the complete reconstruction condition, it also requires the Fourier transform of wavelet $\psi(t)$ satisfies the following "stability condition."

$$A \le \sum_{j=-\infty}^{+\infty} \left|\hat{\psi}(2^{-j}\omega)\right|^2 \le B \tag{7}$$

$0 < A \le B < \infty$.

Definition 2.2

If wavelet $\psi(t)$ satisfy the stability Condition (7), define a "Dual wavelet" $\tilde{\psi}(t)$. Its Fourier transform $\hat{\tilde{\psi}}(t)$ is defined as:

$$\hat{\tilde{\psi}}(\omega) = \frac{\hat{\psi}^*(\omega)}{\sum\limits_{j=-\infty}^{+\infty} \left|\hat{\psi}(2^{-j}\omega)\right|^2} \tag{8}$$

$\hat{\psi}^*(\omega)$ is a conjugate function of $\hat{\psi}(\omega)$.

2.2 Wavelet transform and adaptive time frequency window

Wavelet transform is similar to short-time Fourier transform. The difference is that the wavelet function is used as a window function. Define the time domain window radius of the mother wavelet function $\psi(t)$ of is Δt. The center of the window is t^*, and the frequency domain window radius $\Delta\omega$. The center of the frequency domain window is ω^*, set the time domain window center of $\psi_{a,\tau}(t)$ to $t^*_{a,\tau}$. The center of the time domain window of $\psi_{a,\tau}(t)$ is $t^*_{a,\tau}$. The time domain window radius is $\Delta t_{a,\tau}$. The center of the frequency domain window is $\omega^*_{a,\tau}$. The frequency domain window radius is $\Delta\omega_{a,\tau}$. The center and radius of the time-frequency window of $\psi_{a,\tau}(t)$ and $\psi(t)$ have the following relationship expression:

$$t^*_{a,\tau} = at^* + \tau, \Delta t_{a,\tau} = a\Delta t,$$
$$\omega^*_{a,\tau} = \frac{1}{a}\omega^*, \Delta\omega_{a,\tau} = \frac{1}{a}\Delta\omega$$

It can be seen that the center and width of the time-frequency window of a continuous wavelet $\psi_{a,\tau}(t)$ will expand and contract with the change of scale a.

Wavelet transform uses a time-frequency window to show the time-frequency localization ability of wavelet transform, which is different from short-time Fourier transform. With the change of scale parameter a, the position of the wavelet transform time-frequency window on the phase plane is not only changing but also the shape of the window is changing, as shown in **Figure 1**. **Figure 1(a)** shows that as a time-

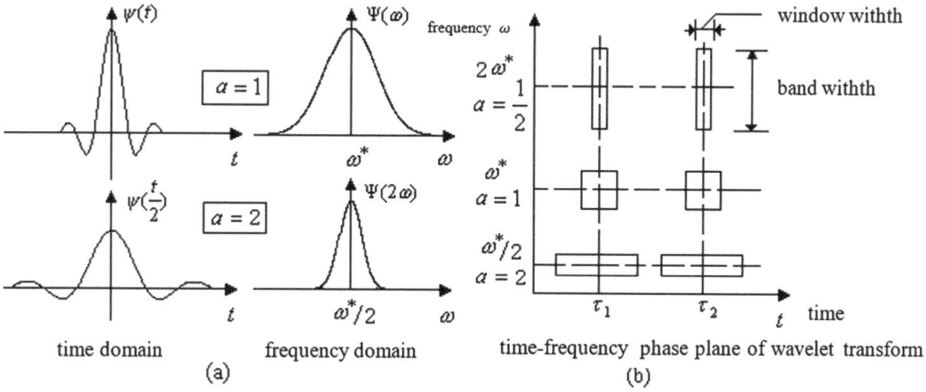

Figure 1.
Schematic diagram of wavelet transform time-frequency window.

frequency window function, when the wavelet function is widened (a increases) in the time domain, its frequency window width is narrowed, and the center of the frequency window is also smaller. **Figure 1(b)** shows that for the same time window center, as the frequency window center (frequency band center) moves up (at which point a decreases), the frequency window width (bandwidth) is widened and the time window width is compressed. Therefore, the shape of the wavelet transform time-frequency window varies with the scale parameter a, which can be analyzed based on the frequency of the signal ω. Adaptive adjustment: At low frequencies, the time resolution of wavelet transform is lower, while the frequency resolution is higher. At high frequencies, the time resolution of wavelet transform is higher, while the frequency resolution is lower. This adaptive characteristic of wavelet transform time-frequency window is also called "auto zoom function," so it is convenient to realize multi-resolution time-frequency analysis for non-stationary signals by using wavelet transform.

Although the center and width of the time window and frequency window of $\psi_{a,\tau}(t)$ vary with a and τ, the area (i.e., window) formed by the time-frequency window on the time-frequency phase plane does not vary with the parameters:

$$\Delta t_{a,\tau}\Delta\omega_{a,\tau} = a\Delta t \cdot \frac{1}{a}\Delta\omega = \Delta t\Delta\omega$$

The area of the time-frequency window of the short-time Fourier transform also has similar properties. This property is called the Heisenberg Uncertainty Theorem, which means the size of Δt and $\Delta\omega$ is mutually constrained, and both cannot be arbitrarily small, so the resolution in both the time and frequency domains cannot be improved without limitation at the same time.

2.3 Typical wavelets

Compared with the standard Fourier transform, the wavelet functions used in wavelet analysis are non-unique, that is, there are various wavelet functions ψ(t) we can apply [6]. The diversity of wavelet analysis is a crucial issue in engineering applications, as selecting the optimal wavelet basis can yield different results when analyzing the same problem using different wavelet bases. At present, the quality of wavelet bases is mainly determined by the error between the results of signal processing using wavelet analysis methods and the theoretical results, thereby determining the quality of wavelet bases.

1. Haar wavelet

$$\psi_H(t) = \begin{cases} 1 & 0 \le t < 1/2 \\ -1 & 1/2 \le t < 1 \\ 0 & else \end{cases} \tag{9}$$

$\psi_H(t)$ is not only orthogonal to $\psi_H(2^j t)\big|_{j \in Z}$ but also orthogonal to its own integer displacement, that is,:

$$\int \psi_H(t)\psi_H(2^j t)\mathrm{dt} = \int \psi_H(t)\psi_H(t-n)\mathrm{dt} = 0, j \in Z, n \in Z \tag{10}$$

2. Daubechies (DbN) wavelet

   Daubechies wavelet is a wavelet function constructed by the world-renowned wavelet analyst Inrid Daubechies. It can generally be abbreviated as Db*N*, where *N* is the order of the wavelet.

   The support region in wavelets $\psi(t)$ and scaling functions $\varphi(t)$ is 2 *N*-1, and the vanishing moment of $\psi(t)$ is *N*. (If $\int t^p \psi(t)\mathrm{dt} = 0, p = 0, 1, \cdots, N-1, N \ge 1$ and $\int t^N \psi(t)\mathrm{dt} \ne t$, the vanishing moment is called *N*.) Except for *N* = 1, DbN does not exhibit symmetry (i.e., nonlinear phase). There is no explicit expression except for *N* = 1.

   The goal of Daubechies wavelets is to construct compactly supported orthogonal wavelets with high-order vanishing moments. The vanishing moment plays an important role in compression, denoising, and singularity detection. When there are singular points (mutation points) in the signal, under high-resolution conditions, the wavelet coefficient at the smooth point is very small, while at the singular point, the wavelet coefficient is large, making it possible to quickly determine the position of the singular point.

3. Symlet (symN) wavelets

4. Morlet wavelets

5. Meyer wavelets

6. Mexican hat(mexh) wavelets

7. Coiflet (coif N) wavelets

8. Biorthogonal (biorNr.Nd) wavelets

9. Reverse Bior wavelets

10. Dmeyer wavelets

11. Gaussian wavelets

12. Complex Gaussian wavelets

13. Complex Morlet wavelets

14. Complex Frequency B-Spline wavelets

15. Complex Shannon wavelets

3. The integrated technology of wavelet analysis and neural network for detecting the average particle size of fluidized bed

The acoustic measurement method for detecting the average particle size of materials in a fluidized bed has the advantages of non-invasive flow field and real-time online detection [7, 8]. However, the acoustic emission signal is a series of time series data. The data is difficult to associate with the chemical parameters in the fluidized bed [9, 10]. Considering that acoustic emission signals are emitted by a large number of particles colliding with each other, or particles colliding with the wall of a fluidized bed, they contain rich multi-scale information. Therefore, the acoustic emission signal can be decomposed at multiple scales, and then, features can be extracted from the information at each scale to form a pattern. Finally, it is associated with the detected object, that is, a certain chemical parameter, to form a soft sensing model for that parameter. There are several issues that need to be addressed. First, decomposition scales of acoustic emission signal, and then how to select and construct mode variables from the information of each scale. Second is how to extract features from these variables when there is a complex correlation between them, and third is how to locate the quantitative relationship between pattern features and the tested object when they are associated. To detect the average particle size of fluidized bed particles by acoustic emission signals, the following methods are proposed in this chapter.

First, multi-scale decomposition of acoustic emission signals is performed using wavelets or wavelet packets to obtain high-frequency and low-frequency detailed signals, energy patterns are constructed based on these signals. Next, principal component analysis (PCA) on the pattern variables to select suitable principal components as feature variables. Finally, establish a neural network model for detecting the average particle size of materials in a fluidized bed.

This chapter eliminates the complex collinearity between variables through multi-scale decomposition of acoustic emission signals and combining PCA screening mode features. When the neural network model is used to detect the average particle size of materials in a cold-model fluidized bed, the detection achieves high accuracy and good stability.

3.1 Artificial neural networks

Artificial neural networks are widely used in fields such as fitting, classification, clustering, feature mining, modeling of control and dynamic systems, and pattern recognition. The feedforward network is the most commonly used and common neural network to date.

People have proposed hundreds of artificial neural network models from different research perspectives, and there are three main types of existing neural networks: namely feedforward neural networks (FFNN), feedback networks (Feedback NN), and self-organizing neural networks (SOMs). In recent years, fuzzy neural networks have also developed rapidly. Fuzzy neural networks organically combine fuzzy technology with neural network technology, and combine the advantages of neural network and fuzzy theory, integrating learning, association, recognition, adaptation, and fuzzy information processing. Fuzzy neural networks have proposed various models and achieved fruitful application results to date. There are mainly feedforward fuzzy neural networks, T-S model fuzzy neural networks, fuzzy maximum minimum neural networks, fuzzy associative memory networks, etc.

3.2 RBF networks and algorithms

The radial basis function neural network (RBFN) was proposed by Powell in 1985 and is essentially a radial basis function method for multivariate interpolation [11, 12]. It is an artificial neural network with simple structure, simple training, wide application, and good generalization ability. It has been widely applied in many fields, especially in the practical applications of function approximation and pattern classification. In terms of structure and operation, RBF networks also belong to feedforward networks, but their neurons have specific settings and learning methods, which make them have specific performance. RBF networks outperform feedforward networks in terms of approximation ability, classification ability, and learning speed. Its advantage lies in using linear learning algorithms to complete the work done by previous nonlinear learning algorithms while maintaining the high accuracy and other characteristics of nonlinear algorithms. Therefore, it is an artificial neural network that has both the fast convergence characteristics of linear algorithms and the high accuracy characteristics of nonlinear algorithms.

Compared with ordinary feedforward networks, RBF networks perform RBF transformations on the input data at the hidden layer. It has been proven mathematically that through RBF transformation, nonlinear separable sample points in one space can be transformed into linearly separable sample points in another space. This is the theoretical basis for the superior performance of RBF networks over ordinary feedforward networks. There are two variants of radial basis neural networks: generalized regression networks (GRNN) and probabilistic neural networks (PNN). The former can be used for function approximation, while the latter can be used for classification.

3.2.1 The structure of RBF networks

The topology of RBF neural networks is a three-layer feedforward network, and the input layer does not perform any transformation on the input information. It only serves the purpose of transmitting data. The kernel function (action function) of hidden layer neurons is a Gaussian function that performs spatial mapping transformation on input information. The third layer is the output layer, which responds to the input mode. The action function of the output layer neurons is a linear function, and the output information of the hidden layer neurons is linearly weighted and output as the output result of the entire neural network. The topology structure of the RBF neural network based on the Gaussian kernel is shown in **Figure 2**. The transfer function of a radial basis function network is based on the distance between the input

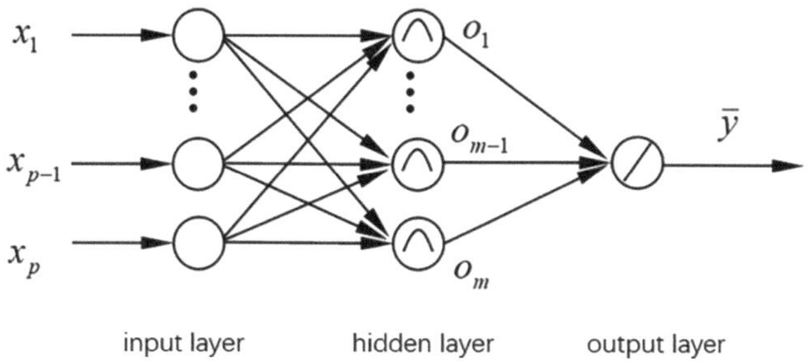

Figure 2.
Topological structure of RBF neural network based on Gaussian kernel.

vector and the threshold vector as the independent variable. When the first input vector is passed to the hidden layer node, the output after processing is:

$$o_j = \exp\left(-\|c_j - x_i\|^2/\sigma_j^2\right) \tag{11}$$

where c_j is the center vector of the Gaussian node of the neuron, and σ_j is the width parameter (spread).

The output layer linearly weights the outputs of each node in the hidden layer as the output of RBFNN, that is, its activation function $y = x$ is a linear function, and the calculated output $\bar{y}_k$ corresponding to the kth input vector x_k is:

$$\bar{y}_k = w_{k0} + \sum_{j=1}^{m} w_{\mathrm{kj}} o_j \tag{12}$$

where w_{kj} is the connection weight, w_{k0} is the bias, and m is the number of nodes in the hidden layer.

3.2.2 Method for selecting RBF neural network centers

The key issue in the learning algorithm of RBF neural networks is the reasonable determination of the central parameters of hidden layer neurons. In existing learning algorithms, the central parameter (or initial value of the central parameter) is either directly selected from a given training sample set using a certain method or determined by clustering methods. Common methods include direct calculation (randomly selecting RBF centers), self-organized learning (selecting RBF centers), supervised learning (selecting RBF centers), and orthogonal least squares (selecting RBF centers).

K-means clustering is a clustering method that clusters according to the minimum distance. The idea of clustering is that for a p-dimensional input vector pattern, it can be seen as points in a p-dimensional **Euclidean** space. If the vectors representing each point are geometrically very close, they can be classified into the same class. Using **Euclidean** distance to measure their proximity:

$$\|x - c\| = \left[\sum_{i=1}^{p} (x_i - c_i)^2\right]^{1/2} \tag{13}$$

where x and c are p-dimensional pattern vector

The main steps for determining the RBF center by the k-means clustering method are given by

① Initialize, set the number of categories k_m, assign initial values to the clustering centers of each category.

② Individual division: Calculate the distance between each individual and each cluster center according to Eq. (13), and divide each individual into different categories based on the principle of minimum distance.

③ Calculate new cluster centers: For the new classes established in step ②, their new center positions can be calculated as

$$c_j(l+1) = \frac{1}{N_j}\sum\nolimits_{x^{(i)} \in S_j(l)} x^{(i)} \quad 1 \le j \le k_m \tag{14}$$

where l is the number of iterations, $c_j(l)$ is the clustering center value of the l-th class j, $x^{(i)}$ is i-th input individual, $S_j(l)$ is the entire class j at l-th iteration, N_j is the number of individuals belonging to class $S_j(l)$ in step ②.

Check convergence: If there is no further change in the clustering center in step ③, convergence has been reached, which satisfies the formula $c_j(l+1) = c_j(l)$, else, back to step ②.

After determining the centers of each cluster, the centers of each radial basis function can be obtained. The extension constant can be set to

$$\text{spread} = \lambda \min_i \left\| c_j - c_i \right\| \tag{15}$$

where λ is the overlap coefficient.

3.2.3 Calculation of BRF network weight parameters

After obtaining the extension constants of each radial basis function center by the K-means clustering algorithm, the second step in the learning process is to use a supervised learning algorithm to obtain the weight of the output layer, the gradient descent method is often adopted.

Define an objective function as:

$$E = \frac{1}{2}\sum_{i=1}^{P} e_i^2 \tag{16}$$

where P is the number of individuals for training samples; e_i is the error signal when inputting the ith individual, defined as:

$$e_i = y_i - \bar{y}_i = y_i - w_{i0} - \sum_{j=1}^{m} w_{ij} \exp\left(-\left\|c_j - x_i\right\|^2 / \sigma_j^2\right) \tag{17}$$

where y_i is the expected output corresponding to the ith input vector x_i, it is a mentor signal.

To minimize the objective function, the correction of each parameter is proportional to its negative gradient, that is,

$$\Delta c_j = -\eta \frac{\partial E}{\partial c_j} \tag{18}$$

$$\Delta\sigma_j = -\eta \frac{\partial E}{\partial \sigma_j} \tag{19}$$

$$\Delta w_j = -\eta \frac{\partial E}{\partial w_j} \tag{20}$$

where (18), (19), (20), η is learning rate.

3.2.4 Learning algorithm of regularized RBF networks

For regularized RBF networks, the number of hidden layer nodes is equal to the number of input-training individuals. All training input individuals are at the center of the radial basis function. Each radial basis function has a uniform width distribution constant. The weight of the output layer is often calculated using the least mean square (LMS) algorithm. The input vector of the LMS algorithm is the output vector of the hidden node. The weight value can be initialized to any value.

3.3 Soft measurement of average particle size in fluidized beds by Sym8 WLA-PCA-MLFN

The multiphase reaction in gas-solid fluidized bed reactors is a typical spatiotemporal and multiscale problem. In order to obtain real, effective, and real-time information on the operation of the bed at the micro-scale, and to control the production process at the macro-scale, it is necessary to establish a correlation mechanism between the micro-system and the macro-system, so that changes in material performance parameters at the micro-scale can be reflected in a timely manner at the macro-scale. It is an important means of establishing such correlations by multi-scale methods. In the production of polyethylene, organic silicon monomer, and granular sodium percarbonate, the particle size and distribution of materials have a certain impact on the reaction rate and product properties. The particle size of materials is also related to reaction time, feeding speed, process parameters, etc., and is often a dynamic and complex process. For example, when conducting the synthesis reaction of dimethyl dichlorosilane in a fluidized bed, it is necessary to timely supplement the raw material with silicon powder and copper powder catalyst. Feeding at the appropriate time can avoid severe fluctuations in the bed temperature, making the process easy to control. The particle size and distribution of the material should be kept stable. If the particle size distribution is not reasonable, it may affect the reaction effect. It is difficult to conduct real-time online detection of material particle size in fluidized beds. As a new measurement method, acoustic measurement has the characteristic of real-time response to changes in reactor material particle size or concentration. It can be considered to use acoustic emission signals for online detection of material characteristics in the reactor. Acoustic emission signal detection has undergone rapid development since its application in the 1980s. But acoustic emission signals are a type of wave, often recorded as a series of temporal data points, and have sudden transients, often mixed with interference noise. How to effectively extract correlation mode information from acoustic emission signals is an urgent problem to be solved. The acoustic emission signal is related to various factors such as bed height, material composition, temperature, and empty bed gas velocity. For a stable fluidized bed, the acoustic emission signal is mainly influenced by the particle size and distribution of the material. Spectrum analysis, wavelet analysis, wavelet packet analysis, fractal feature analysis, or complexity analysis can all be used to analyze acoustic emission signals. Among them, wavelet analysis decomposes acoustic emission signals into low-frequency overview signals and high-frequency detail signals, which have significant advantages for analyzing nonlinear, non-stationary pulsating signals. Previous literature often focused on the spectrum of acoustic emission signals, which is relatively complex. The models established using this method are generally only suitable for qualitative analysis. Lack of quantitative indicators: In the construction of association patterns and feature selection, more concise methods can be considered.

The methods used in this section are as follows: First, wavelet analysis (WLA) is performed on the collected acoustic emission signals. Calculate the energy mode of the acoustic emission signal from the decomposed low-frequency profile signal and high-frequency detail signal. In order to eliminate the multicollinearity between variables, principal component analysis (PCA) was performed on the energy pattern, and principal components were extracted from the energy pattern. Then, using the principal component as the input and the average particle size in the fluidized bed as the output, a multi-layer feed-forward neural network (MLFN) is constructed to establish the quantitative relationship between the principal component of the acoustic emission signal energy mode and the average particle size. The experimental results show that when using this method to predict the average particle size of materials in the bed, the computational cost is not large and the accuracy is high.

3.3.1 The relationship between acoustic emission signals and fluidized bed particle size

There is a large amount of collision and friction between particles in the fluidized bed, as well as between particles and the container wall. The impact force generated by particles of different particle sizes when impacting the container wall or colliding with each other is significantly different, which will generate strong or weak acoustic emission signals with different frequencies and transmit them outward through the container wall and air in the form of elastic waves. Under the same other process conditions, the acoustic emission signals generated by particle groups of different particle sizes are different. Thus, acoustic emission signals can be used to detect the particle size and distribution of the fluidized bed.

For n rigid spherical particles with particle size d_p and mass m, the force generated by impact on the wall surface with area ΔA is given as [13]:

$$F(t) = \sum_{i=1}^{n} 2mu_i\delta(t - t_i) \tag{21}$$

where t is time, $\delta(t)$ is Dirac delta function related to time t, t_i is the time for the i-th particle to reach the wall, u_i is the velocity at which the u-th particle vertically impacts the wall surface. There are $f_p \cdot T$ impacts between particles and the wall surface within time T, f_p is the average arrival rate of particles on the area ΔA, that is, the frequency of particles hitting the wall. Therefore, the average force per unit time of these particles is:

$$\overline{F(t)} = \frac{\int_0^T F(t)dt}{T} = \frac{2mu\int_0^T \sum_{i=1}^{n} \delta(t - t_i)dt}{T} \tag{22}$$

where u is the average velocity of particles impacting the wall vertically, $\int_0^T \sum_{i=1}^n \delta(t - t_i)dt = f_p T$, therefore:

$$\bar{F(t)} = 2muf_p \tag{23}$$

The resulting acoustic emission pressure is:

$$P_a = \frac{\eta \cdot \overline{F(t)}}{\Delta A} \tag{24}$$

In the equation, η is the efficiency of converting impact pressure into sound pressure. Assuming that the concentration of particles colliding with the wall near the wall is C (pieces/m^3), which is inversely proportional to the square of the projected diameter of the particles d_p^2 on the wall, that is:

$$C = \xi / d_p^2 \tag{25}$$

in the equation, ξ is the proportional coefficient, and the frequency of particle impact on the wall is:

$$f_p = C \cdot \frac{\Delta A v}{\Delta A} = C \cdot u \tag{26}$$

Thus, the average AE flux of particles with particle size d_p and mass m impacting the wall surface per unit of time is

$$J = P_a \Delta A u = 2\xi\eta m u^3 / d_p^2 \tag{27}$$

The AE energy of particles with particle size d_p and mass m colliding with the wall at duration T is

$$E = \int_0^T J dt = \int_0^T \frac{\pi}{3} \xi \eta \rho_s u^3 d_p dt \tag{28}$$

It can be seen that AE energy is a function of particle size d_p, material density ρ_s, and duration T. Since particles with different particle sizes have different AE energy, within a certain duration T, while keeping other parameters constant, changes in the average particle size and its distribution can be understood through acoustic emission signals,

In a fluidized bed reactor, the average particle size often changes continuously with the progress of the reaction process. The system maintains a dynamic balance. In the normal state, the characteristic mode of the acoustic emission signal will not undergo significant fluctuations. Once the balance of the system is disrupted, such as by fluctuations in particle size, agglomeration of materials in the reactor, the acoustic emission signal will also change accordingly.

Assuming that there are J types of different particle sizes acting together on the wall, the percentage of particles with particle size $d_{p,j}$ is x_j. According to the principle of linear superposition of acoustic energy, the relationship between the acoustic energy E_j generated by particles with particle size $d_{p,j}$ and the total sound energy is:

$$\sum_{j=1}^{N} \frac{E(d_{p,j})}{E_{\text{mix}}} x_j = 1 \tag{29}$$

where, $E_{\text{mix}} = \sum_{j=1}^{N} E(d_{p,j}) x_j$,

That is to say, the AE signals generated by mixed particle sizes can be seen as the sum of various acoustic emission signals with different particle sizes. Therefore, based on the AE signals, the average particle size and particle size distribution of particle groups under microscale conditions can be predicted.

3.3.2 Multi-scale decomposition of acoustic emission signals

The raw acoustic emission signals collected under specific process conditions can be regarded as a wide, stationary random time series. A complete description requires analysis from three aspects: amplitude domain, time domain, and frequency domain to extract features. This chapter first performs multi-scale decomposition of acoustic emission signals, then calculates the total energy of each detail signal, and extracts principal components as pattern features.

Assuming there is an acoustic emission signal sequence: $\left\{c_1^j, c_2^j, \cdots, c_k^j, \cdots c_n^j\right\}$, where the superscript j represents the observation scale, and the subscript represents the serial number, when $j = 0$ it represents the original acoustic signal. At resolution $2^{-j}(j \in Z$, the discrete approximation of the signal can be represented as:

$$f_j(t) = \sum_{j,k \in Z} c_k^j \varphi_{j,k}(t) \tag{30}$$

$\varphi_{j,k}(t)$ is the scale function series under the resolution of a condition 2^{-j}

According to the theory of wavelet decomposition, signals can be decomposed at multiple scales to calculate their approximate $\hat{A}_{j+1}f$ and detailed information $\hat{D}_{j+1}f$ under higher resolution conditions:

$$f_j(t) = \hat{A}_{j+1}f + \hat{D}_{j+1}f = \sum_k c_k^{j+1} \varphi_{j+1,k}(t) + \sum_k d_k^{j+1} \psi_{j+1,k}(t) \tag{31}$$

where d_k^{j+1} is wavelet coefficients on the $j+1$ scale, $\psi(t)$ is wavelet function, $\hat{A}, \hat{D}$ is an operator for calculating low-frequency detail signals and high-frequency detail signals. Because of translational invariability and the orthogonality of expansion and contraction of φ and ψ, it can be calculated that

$$c_k^{j+1} = \sum_n c_n^j \langle \phi_{j,n}, \phi_{j+1,k} \rangle = \sum_n c_n^j h_{n-2k}^* \tag{32}$$

Similarly,

$$d_k^{j+1} = \sum_n c_n^j \langle \phi_{j,n}, \psi_{j+1,k} \rangle = \sum_n c_n^j g_{n-2k}^* \tag{33}$$

Eqs. (32) and (33) are decomposition algorithms for wavelets. Where "*" represents the conjugation of complex numbers, $\{h_k\}_{k \in Z}$. It is the sequence of filter coefficients corresponding to the two scale equations of the orthogonal scaling function, which can be regarded as a low-pass filter, $\{g_k\}_{k\in Z}$ can be regarded as a high-pass filter. From the wavelet coefficients and scale coefficients at the $j+1$ scale, the scale coefficients at the j scale can be obtained through reconstruction algorithms:

$$\begin{aligned} c_k^j = \sum\nolimits_n c_n^{j+1}\langle \varphi_{j+1,n}, \varphi_{j,k}\rangle + \sum\nolimits_n d_n^{j+1}\langle \psi_{j+1,n}\varphi_{j,k}\rangle \\ \sum\nolimits_n c_n^{j+1} h_{n-2k} + \sum\nolimits_n d_n^{j+1} g_{k-2n} \end{aligned} \tag{34}$$

The amplitude of the acoustic emission signal can also be regarded as energy, and when other process conditions are constant, it is mainly influenced by the particle size. Therefore, the energy value of the detail signal can reflect the size of the particle. For mixed-particle systems with different particle sizes, this difference can be reflected in detailed information through multi-scale decomposition of acoustic emission signals. The energy of the low-frequency profile and high-frequency detail signal after wavelet decomposition of the acoustic emission signal at the defined fs resolution can be represented by the following equation:

$$\varepsilon_a^{j+1} = \sum\nolimits_k \left| c_k^{j+1} \right| \tag{35}$$

$$\varepsilon_d^{j+1} = \sum\nolimits_k \left| d_k^{j+1} \right| \tag{36}$$

It can be seen from the multi-resolution that the sum of the scale coefficients of low-frequency signals and the absolute values of wavelet coefficients of high-frequency detail signals can be calculated as the energy mode of acoustic emission signals.

$$f_j = f_{j+1} + d_{j+1} = f_{j+2} + d_{j+2} + d_{j+1} = \cdots = f_p + d_p + d_{p-1} + \cdots + d_{j+1}$$

When the decomposition scale of the acoustic emission signal is p, a $p+1$ dimensional energy mode can be obtained: $\left(\varepsilon_a^p, \varepsilon_d^1, \varepsilon_d^2, \cdots, \varepsilon_d^p\right)$.

A certain quantitative relationship can be established between the average particle size and the energy mode of acoustic emission signals as follows:

$$y = y\left(\varepsilon^j\right) = y\left(\varepsilon_a^p, \varepsilon_d^1, \varepsilon_d^2, \cdots, \varepsilon_d^p\right) \tag{37}$$

In the equation above, the subscript a represents low frequency and the subscript d represents high frequency. P represents the maximum scale of discrete signal decomposition with a length of 2^n, and $p \le n-1$. The sample data matrix composed of the energy mode of the m observation of the acoustic emission signals generated by particles in a certain interval in a fluidized bed is:

$$X = \begin{pmatrix} \varepsilon_{a,1}^p & \varepsilon_{d,1}^1 & \cdots & \varepsilon_{d,1}^p \\ \varepsilon_{a,2}^p & \varepsilon_{d,2}^1 & \cdots & \varepsilon_{d,2}^p \\ \vdots & \vdots & \vdots & \vdots \\ \varepsilon_{a,m}^p & \varepsilon_{d,m}^1 & \cdots & \varepsilon_{d,m}^p \end{pmatrix}_{m\times(p+1)} \tag{38}$$

Considering the significant differences in signal energy among different details, it is usually necessary to standardize the standard deviation of the above energy mode matrix before use.

After p-scale decomposition of acoustic emission signals, the energy mode has a total of $p + 1$ components, and there is often a complex collinearity between them. The energy mode is directly associated with the detection object, which not only has a large number of independent variables but may also cause interference with each other. To eliminate multicollinearity, methods such as principal component analysis can be used. At present, many data processing in process control and monitoring also follow this model, showing good application prospects.

3.4 Experiment and discussion

3.4.1 Collection of fluidized bed acoustic emission signals

The fluidized bed acoustic emission signal and data acquisition and analysis system (UNIAE2003) was developed by Professor Yang Yongrong's research group at the Joint Chemical Reaction Engineering Research Institute of Zhejiang University. The data acquisition and analysis system includes acoustic sensors (AE sensors), amplifiers, A/D conversion cards, and computers, which can achieve multi-channel data acquisition. The device process is shown in **Figure 3**, and the sensors are tightly attached to the outer wall of the cold model fluidized bed, which can basically ignore other external noise. The sampling frequency of the data collection system is 500 kHz. To avoid interference and reduce the impact of external noise, hardware filters were used for sampling signals for differential filtering preprocessing.

The material used in the fluidized bed is granular polyethylene with different particle sizes. The operating gas speed is 0.6 m/s. The particles are divided into five particle size intervals: 1 ~ 0.90, 0.90 ~ 0.60, 0.60 ~ 0.45, 0.45 ~ 0.22, and 0.22 ~ 0.180 mm. Twenty sampling observations are conducted on the acoustic emission signals of each particle size interval, and the length of the data recording points obtained from each observation is 16,384. This is the number of data points recorded by the recorder. Acoustic emission signals can also be collected under different bed heights, materials, and temperature conditions by this experimental equipment.

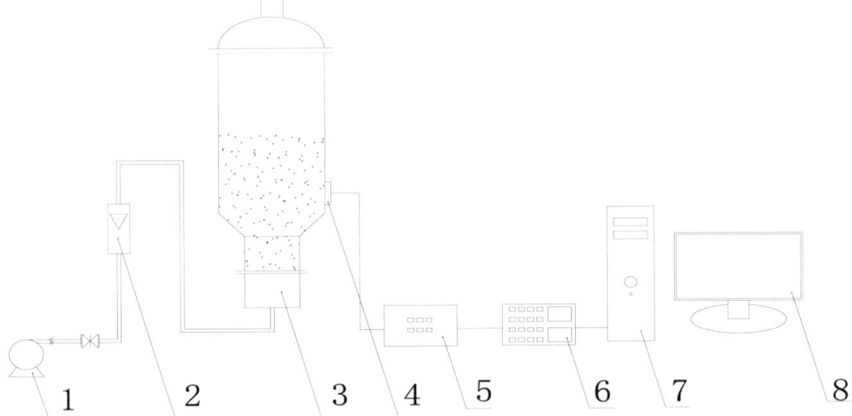

Figure 3.
Schematic diagram of fluidized bed acoustic emission signal data processing system. 1-blower, 2-gas flow meter, 3-fluidized. Bed, 4-acoustic emission signal sensor, 5-signal amplifier, 6-signal processor, 7-computer host, 8-monitor.

Multi-scale decomposition of the acoustic emission signal recorded by the instrument is used to construct the energy mode of the acoustic emission signal, and the wavelet types and decomposition scales should be screened. Perform principal component analysis on the energy mode components and select several principal components. Strive for good detection results.

3.4.2 Energy mode and principal component analysis of acoustic emission signals

The original acoustic emission signal sequence is processed according to the following steps:

3.4.2.1 Select decomposition wavelet

Many wavelets or wavelet packets that can be selected to decompose AE signals so as to obtain energy patterns. There is no clear regulation on which wavelet is more suitable for decomposing them. When selecting wavelet types, the minimum error generated by signal decomposition reconstruction can be taken as the indicator. The best decomposed wavelet (or wavelet packet) should have the minimum reconstruction error.

The original acoustic emission signal with a length of 16,384 is decomposed at 6 scales using different wavelets, and then reconstructed using the same wavelet. The absolute error sum is used as the judgment indicator, as shown in Eq. (39):

$$\mathrm{Err}_w = \sum_{i=1}^{n} \left| C_i^0 - \tilde{C}_i \right| \tag{39}$$

where $n = 16384$, C_i^0 is the original sequence of acoustic emission signals, $\tilde{C}_i$ is a reconstructed sequence of acoustic emission signals. The subscript i represents the serial number of the acoustic emission signal. Examine the **Haar** wavelet, **Daubechies** wavelet series, and **Sym** wavelet series. It was found that the reconstruction error of **Sym8** wavelet is the smallest. Therefore, **Sym8** wavelet is chosen as the wavelet for decomposing acoustic emission signals.

3.4.2.2 Energy mode construction and preprocessing

A 9-dimensional energy pattern was obtained by Sym8 wavelet 8-scale decomposition on the acoustic emission signal. **Table 1** shows the energy patterns obtained at five different particle sizes (partial). From the data in the table, it can be seen that the energy of high-frequency detail signals is much greater than that of low-frequency detail signals, with a difference of one order of magnitude. At this point, logarithms can be taken for the energy mode data so that the data is on the same order of magnitude.

Due to the fact that both high-frequency and low-frequency detail signals are decomposed from the original acoustic emission signal, there exists a strong multicollinearity. Principal component analysis of the data is required.

The purpose of principal component analysis is to convert multiple existing indicators into fewer linearly unrelated comprehensive indicators. For a p-dimensional random vector $x = (x_1, x_2, \cdots, x_p)$, if there is a complex correlation between its variables, they can be summarized by mk$(mk \le p)$ "comprehensive variables" or "principal components," which are linear combinations of the original p variables.

	$\Sigma\lvert d_1\rvert$	$\Sigma\lvert d_2\rvert$	$\Sigma\lvert d_3\rvert$	$\Sigma\lvert d_4\rvert$	$\Sigma\lvert d_5\rvert$	$\Sigma\lvert d_6\rvert$	$\Sigma\lvert d_7\rvert$	$\Sigma\lvert d_8\rvert$	$\Sigma\lvert a_8\rvert$
1 ~ 0.90 mm	1.2271	1.1938	1.1605	1.1500	1.1003	0.6694	0.1671	0.0471	0.0471
	1.2926	1.2603	1.2318	1.2209	1.1588	0.7235	0.2219	0.0885	0.0885
	⋮	⋮	⋮	⋮	⋮	⋮	⋮	⋮	⋮
0.90 ~ 0.60 mm	1.0690	1.0516	1.0250	1.0090	0.9508	0.6064	0.2298	0.0602	0.0602
	1.1206	1.0988	1.0695	1.0542	1.0058	0.6553	0.2370	0.0634	0.0634
	⋮	⋮	⋮	⋮	⋮	⋮	⋮	⋮	⋮
0.60 ~ 0.45 mm	1.4479	1.2481	0.8552	0.6495	0.4915	0.3104	0.1160	0.0505	0.0505
	1.5587	1.3527	0.9110	0.6915	0.4862	0.2869	0.1199	0.0544	0.0544
	⋮	⋮	⋮	⋮	⋮	⋮	⋮	⋮	⋮
0.45 ~ 0.22 m	0.2171	0.2007	0.1788	0.1535	0.1083	0.0555	0.0483	0.0488	0.0488
	0.2218	0.2040	0.1771	0.1444	0.1061	0.0549	0.0477	0.0481	0.0481
	⋮	⋮	⋮	⋮	⋮	⋮	⋮	⋮	⋮
0.22 ~ 0.180 mm	0.1487	0.1267	0.1179	0.1126	0.1026	0.0453	0.0404	0.0402	0.0402
	0.1398	0.1182	0.1078	0.1014	0.0902	0.0514	0.0440	0.0435	0.0435
	⋮	⋮	⋮	⋮	⋮	⋮	⋮	⋮	⋮

Table 1.
Energy patterns obtained from 8-scale decomposition of acoustic emission signals($\times 1.0\ e^{-3}$).

Under the principle of less loss of useful information, replace more original multidimensional variables with fewer comprehensive variables to achieve dimensionality reduction of high-dimensional data. The principle of selecting the number of principal components usually ensures that the sum of the variance contribution ratio (SVCR) reaches or exceeds 90%, as shown in Eq. (40). SVCR expresses the amount of information that *mk* principal components account for all features of $x_1, x_2, \cdots, x_p$. **Table 2** lists the results of principal component analysis of the energy mode of acoustic emission signals after 8-scale decomposition. The variance occupied by the following three principal components is 0.0017%, so it is not listed.

$$\text{SVCR} = \sum_{i=1}^{\text{mk}} \lambda_i / \sum_{i=1}^{p} \lambda_i = \sum_{i=1}^{\text{mk}} \lambda_i / p \tag{40}$$

In the equation, λ_i is the eigenvalues of the sample covariance matrix, $\lambda_1 \geq \lambda_2 \geq \cdots \geq \lambda_p \geq 0$

3.4.2.3 Determine decomposition scale

Acoustic emission signals of different size particles were multi-scale decomposed by Sym8 wavelet, and then, the energy of the corresponding scale acoustic emission signals was obtained. After range normalization processing and principal component analysis, the variance of each principal component is shown in **Table 3**. It can be seen that the ratio of the maximum variance to the minimum variance of the principal components is quite large, indicating that multicollinearity is very serious. Considering that the excessively high decomposition scale leads to an increase in computational

Particle size/principal component	f_1	f_2	f_3	f_4	f_5	f_6
1 ~ 0.90 mm	−1.7612	−1.3737	−1.3862	−0.5974	−0.0494	0.0086
	−3.8690	1.0179	0.3630	−0.1481	0.0058	−0.0074
	⋮	⋮	⋮	⋮	⋮	⋮
0.90 ~ 0.60 mm	−1.9344	−0.2136	−0.9412	0.2943	0.0443	−0.0063
	−2.3201	−0.0994	−0.8812	0.2670	−0.0128	0.0113
	⋮	⋮	⋮	⋮	⋮	⋮
0.60 ~ 0.45 mm	−0.2694	−1.4987	0.3921	0.1166	−0.0365	0.0284
	−0.5988	−1.4909	0.7059	0.1738	0.0594	−0.0148
	⋮	⋮	⋮	⋮	⋮	⋮
0.45 ~ 0.22 m	2.9133	0.5259	0.1801	0.0004	0.0319	−0.0333
	2.9517	0.4738	0.1563	0.0036	0.0229	−0.0244
	⋮	⋮	⋮	⋮	⋮	⋮
0.22 ~ 0.18 mm	3.4536	0.1576	−0.2595	−0.0546	0.0078	0.0263
	3.3436	0.3753	−0.1152	−0.0024	−0.0217	0.0232
	⋮	⋮	⋮	⋮	⋮	⋮

Table 2.
Principal components of energy patterns obtained from scale decomposition (partial).

Decomposition scale	3	4	5	6	7	8
λ_1	3.8709	4.7243	5.55	6.4723	7.2595	7.7735
λ_2	0.12901	0.27542	0.44917	0.51595	0.59732	0.83072
λ_3	8.051e-5	0.0001945	0.000689	0.011085	0.13799	0.29968
λ_4	1.87e-35	5.054e-5	0.000117	0.0005371	0.0045857	0.090915
λ_5		3.85e-35	4.817e-5	0.0001136	0.00049631	0.0045248
λ_6			3.59e-36	4.4803e-5	0.000111	0.0004932
λ_7				1.046e-34	4.3571e-5	0.0001101
λ_8					5.4742e-32	4.269e-5
λ_9						1.475e-31

Table 3.
Variance of principal components under different decomposition scales.

time. The decomposition scale is too small, and it is difficult to grasp the rich information contained in each detail. Therefore, the decomposition scale is determined to be 6, resulting in a 7-dimensional energy pattern. Among the seven principal components, the first three principal components account for 99.99% of the total variance, which is sufficient to extract all information. Therefore, the first three principal components are selected as inputs for the neural network.

3.4.2.4 Construction of neural network model

During each experiment, among 100 individuals, 18 individuals were selected from samples of each granularity (calculated principal components) as training sample data, 1 as validation data, and 1 as data for the test network. A total of 90 individuals were used for training, 5 for validation, and 5 for testing. Adopting a cross-over approach. The purpose of validation is to prevent the network from being overtrained (overfitting). During the experiment, the three principal components extracted were used as inputs to the network, and the average particle size was used as the output of the network. The number of output neurons is 1, and the number of input neurons is 3. The network has three layers, and the hidden layer transfer function is "tansig," which is a hyperbolic tangent type transfer function. The output layer transfer function is "purelin," which is a linear transformation function. The L-M (Levenberg Marquardt) algorithm is used for network training, with the aim that the L-M algorithm does not need to calculate the Hessian matrix and has a fast convergence speed. The target error during training is set to 10^{-3}. According to the principle that the number of hidden layer nodes is roughly twice the number of input layer nodes [14], this chapter determines that the number of hidden layer nodes is 6, so the network structure is 3-6-1. The stopping condition for training is that the training error is less than the set target value.

The error function used for validation is specified as follows:

As for acoustic emission signals generated at specific particle sizes, compare the predicted particle size $\hat{y}_i$ calculated by the network with the actual average particle size y_i for the acoustic emission signals generated at specific particle sizes. Calculate the mean square error

$$\text{RMSE} = \sqrt{\frac{1}{N}\sum_{i=1}^{N}(y_i - \hat{y}_i)^2} \tag{41}$$

N is the number of individuals used for validation.

4. Results and discussion

Fitting error refers to randomly selecting five individuals (one for each granularity) from the training sample, using the trained network for prediction, and calculating the absolute value of relative error. Prediction error is the absolute value of the relative error obtained when using a trained network to predict test samples. Each particle size has 20 individuals, and 20 fitting and prediction experiments are conducted. The results are listed in **Table 4**. From **Table 4**, it can be seen that for larger particles, the fitting and prediction accuracy are relatively high when predicting particle size based on principal component analysis. However, for small particles, the prediction accuracy is relatively low. It is possible that the signal-to-noise ratio of small particles is slightly lower than that of large particles. The average prediction accuracy of all acoustic emission signal samples is 94.25%, and the average fitting accuracy is 97.7%. This situation indicates that in actual production processes, the average particle size of materials in the bed can be analyzed and predicted online from the acoustic emission signals emitted by the fluidized bed.

Particle size range /mm		**1 ~ 0.90 mm**		**0.90 ~ 0.60 mm**		**0.60 ~ 0.45 mm**		**0.45 ~ 0.22 mm**		**0.22 ~ 0.180 mm**	
Sample number		fitting	prediction	fitting	prediction	fitting	prediction	fitting	prediction	fitting	prediction
Fitting and prediction accuracy	1	0.98635	0.97046	0.98634	0.96996	0.9798	0.97795	0.98102	0.9129	0.9833	0.89281
	2	0.94695	0.90697	0.98911	0.97954	0.97596	0.98421	0.98099	0.89347	0.98865	0.89233
	3	0.98677	0.97687	0.98711	0.98291	0.97701	0.98174	0.98496	0.90512	0.98647	0.89963
	4	0.96576	0.97951	0.95559	0.9701	0.97583	0.97254	0.98137	0.90389	0.98075	0.90041
	5	0.98868	0.93139	0.96554	0.98004	0.97788	0.96411	0.97755	0.89583	0.97233	0.89578
	6	0.97757	0.97436	0.98926	0.97579	0.97662	0.97618	0.98935	0.89707	0.97311	0.89718
	7	0.98867	0.97806	0.92952	0.97207	0.97772	0.97132	0.98367	0.90033	0.98327	0.89986
	8	0.93635	0.98478	0.98913	0.98287	0.978	0.9661	0.97717	0.89994	0.98195	0.89378
	9	0.98555	0.98633	0.95537	0.97242	0.97688	0.97724	0.98303	0.89962	0.98296	0.90367
	10	0.97805	0.97313	0.98672	0.98114	0.97833	0.97791	0.98428	0.89221	0.97792	0.9003
	11	0.98705	0.98619	0.98674	0.98292	0.98046	0.97928	0.97333	0.90851	0.98614	0.90075
	12	0.98602	0.9242	0.9874	0.96456	0.97779	0.97367	0.89982	0.89766	0.98604	0.90798
	13	0.96688	0.8971	0.97626	0.98975	0.97698	0.97291	0.97592	0.90047	0.9814	0.89611
	14	0.98439	0.97782	0.98573	0.98178	0.97837	0.97397	0.99054	0.90144	0.98879	0.90775
	15	0.98841	0.97475	0.98642	0.98025	0.97708	0.97413	0.98429	0.9046	0.97678	0.90528
	16	0.98572	0.98082	0.98841	0.97116	0.97534	0.97609	0.94056	0.90255	0.97972	0.89917
	17	0.9906	0.97829	0.99008	0.97371	0.97773	0.98342	0.9878	0.90123	0.98374	0.90157
	18	0.98874	0.96725	0.9873	0.96814	0.97699	0.9756	0.98739	0.893	0.97851	0.9071
	19	0.96582	0.95985	0.98665	0.96866	0.97659	0.97825	0.9816	0.90485	0.92117	0.90164
	20	0.97635	0.9077	0.97634	0.97053	0.9798	0.9554	0.96642	0.90797	0.90247	0.90238
mean value		0.97803	0.960792	0.979251	0.97592	0.97755	0.97460	0.97555	0.901133	0.974774	0.900274

Table 4.
The fitting and prediction accuracy of Sym8-WLA-PCA-MLFN for particle size.

Author details

Ximing Chen
Chemistry and Material Science College, Huaibei Normal University, Anhui, China

*Address all correspondence to: chenxm@chnu.edu.cn

References

[1] Sun Y. Wavelet Analysis and Application. Beijing, China: China Machine Press; 2005

[2] Burke B. The mathematical microscope: Waves, wavelets, and beyond. In: Bartusiak M, et al., editor. Apositron Named Priscilla, Scientific Discovery at the Frontier, Chapter 7. Washington DC: National Academy Press; 1994. pp. 196-235

[3] Akansu AN, Smith MJT. Subband and Wavelet Transforms, Design and Applications. Boston: Kluwer Academic Publishers; 1996

[4] Beylkin G, Coifman RR, Rokhlin V. Fast wavelet transforms and numerical algoritms I. Communications on Pure and Applied Mathematics. 1991;**44**:141-183

[5] Ten DI. Ten Lectures on Wavelets. Philadelphia, PA: SIAM; 1992

[6] Feisi Technology Product Research and Development Center. Wavelet Analysis Theory and Matlab 7 Implementation. Beijing, China: Publishing House of Electronics Industry; 2005

[7] Belchamber RM, Betteridge D, Collihs MP, et al. Quantitative study of acoustic emission from a model chemical process. Analytical Chemistry. 1986;**58**: 7873-7877

[8] Wade AP, Sibbal DB, Bailey MN, et al. An analyticla perspective on acoustic emission. Analytical Chemistry. 1991;**63** (9):497-507

[9] Wentzell PD, Wade AP. Chemical acoustic emission analysis in the frequency domain. Analytical Chemistry. 1989;**61**(23):2638-2642

[10] Hansmann H. Application of acoustic emission analysis on adhesion and structural problems of organic and metallic coatings. Industrial & Engineering Chemistry Product Research and Development. 1985;**24**(2): 252-257

[11] Powell MJD. Radial basis functions for multivariable interpolation: A review. In: Mason JC, Cox MG, editors. Algorithms for Approximation. Oxford; 1987. pp. 143-167

[12] Powell MJD. Radial basis function approximations to polynomials. In: Proceedings of the 12th Biennial Numerical Analysis Conference, Dundee. 1987. pp. 223-241

[13] Linxi H. Research on multiscale structures of acoustic measurement and fluidized bed polymerization reactors, Doctoral Dissertation of Zhejiang University. Hangzhou, China: Zhejiang University

[14] Berger J, Coifman RR, Goldberg MJ. Removing noise from music using local trigonometric bases and wavelet packets. Journal of the Audio Engineering Society. 1994;**42**(10):808-817

Chapter 2

An Optimization Approach to Supervised Principal Component Analysis

Anthony O. Smith and Anand Rangarajan

Abstract

Supervised dimensionality reduction has become an important theme in the last two decades. Despite the plethora of models and formulations, there is a lack of a simple model that aims to project the set of patterns into a space defined by the classes (or categories). We set up a model where each class is represented as a 1D subspace of the vector space formed by the features. Assuming the set of classes does not exceed the cardinality of the features, the model results in multi-class supervised learning in which the features of each class are projected into the class subspace. Class discrimination is guaranteed via the imposition of the orthogonality of the 1D class sub-spaces. The resulting optimization problem—formulated as the minimization of a sum of quadratic functions on a Stiefel manifold—while being non-convex (due to the constraints), has a structure for which we can identify when we have reached a global minimum. After formulating a version with standard inner products, we extend the formulation to a reproducing kernel Hilbert space and similarly to the kernel version. Comparisons with the multi-class Fisher discriminants and principal component analysis showcase the relative merits toward dimensionality reduction.

Keywords: dimensionality reduction, optimization, classification, supervised learning, Stiefel manifold, category space, Fisher discriminants, principal component analysis, multi-class

1. Introduction

Dimensionality reduction and supervised learning have long been active tropes in machine learning. For example, principal component analysis (PCA) and the support vector machine (SVM) are standard bearers for dimensionality reduction and supervised learning. Even now, machine learning researchers are accustomed to performing PCA when seeking a simple dimensionality reduction technique, even though it is an unsupervised learning approach. In the past decade, there has been considerable interest in including supervision (expert label information) in dimensionality reduction techniques. Beginning with the well-known EigenFaces versus FisherFaces debate

[1], considerable activity has centered around using Fisher linear discriminants (FLD) and other supervised learning approaches in dimensionality reduction. Since the Fisher linear discriminant has a multi-class extension, it is natural to begin there. However, asking if this is the only possible approach is also natural. In this work, with a fundamental goal, we design a category space approach using multi-class information to reduce dimensionality.

The venerable Fisher discriminant is a supervised dimensionality reduction technique wherein a maximally discriminative one-dimensional subspace is estimated from the data. The criterion used for discrimination is the ratio between the squared distance of the projected class means and a weighted sum of the projected variances. This criterion has a closed-form solution yielding the best 1D subspace. The Fisher discriminant also has an extension to the multi-class case. Here, the criterion used is more complex and highly unusual: the squared distance ratio between each class's projected mean compared to the total projected mean and the sum of the projected variances. This, too, results in a closed-form solution but with the subspace dimension cardinality being one less than the number of classes.

The above description of the multi-class FLD sets the stage for our approach. We assume the set of categories (classes) is a subspace of the original feature space (similar to FLD). However, we add the restriction that the category bases are mutually orthogonal with the origin of the vector space belonging to no category. Given this restriction, the multi-class category space dimensionality reduction criterion is quite straightforward. We maximize the square of the inner product between each pattern and its category axis to discover the category space via this process. (Setting the origin is a highly technical issue and, therefore, not described here.) The result is a sum of quadratic objective functions on a Stiefel manifold—the category space of orthonormal basis vectors. This very interesting objective function has coincidentally received quite a bit of treatment recently [2–5]. Furthermore, there is no need to restrict ourselves to sums of quadratic objective functions provided we are willing to forego useful analysis of this base case. The unusual aspect of the objective function comprising sums of quadratic objective functions is that we can formulate a criterion that guarantees that we have reached a global minimum if the achieved solution satisfies it.

While numerous alternatives exist to the FLD (such as canonical correlation analysis [6, 7]) and while there are many nonlinear unsupervised dimensionality reduction techniques (such as local linear embedding [8–10], ISOMAP [11, 12] and Laplacian Eigenmaps [13–15]), we have not encountered a simple dimensionality reduction technique which is based on projecting the data into a space spanned by the categories. Unfortunately, no algorithm at present can *a priori* guarantee satisfaction of this criterion; hence, we can only check on a case-by-case basis. Despite this, our experimental results show that we get efficient solutions, competitive with those obtained from other dimensionality reduction algorithms.

2. Related work

Traditional dimensionality reduction techniques like principal component analysis (PCA) [16–18] and supervised algorithms such as Fisher linear discriminant analysis [19] seek to retain significant features while removing insignificant, redundant, or noisy features. Many real-world problems have been solved using these algorithms as preprocessing steps before applying a classification algorithm. A limitation of most methods is that there is no specific connection between the dimensionality reduction technique and the supervised learning-driven classifier. Dimensionality reduction techniques such as canonical correlation analysis (CCA) [20], and partial least squares (PLS) [21, 22] on the one hand and classification algorithms such as support vector machines (SVM) [23] on the other seek to optimize different criteria. In contrast, in this paper, we analyze dimensionality reduction from the perspective of multi-class classification. Using a category vector space (with dimension equal to class cardinality) is an integral aspect of this approach.

In supervised learning, it is customary for classification methodologies to regard classes as nominal labels without having any internal structure. This remains true regardless of whether a discriminant or classifier is sought. Discriminants are designed by attempting to separate patterns into opposing classes [24–26]. When generalization to a multi-class classifier is required, many oppositional discriminants are combined, with the final classifier being a winner-take-all (or voting-based) decision w.r.t. the set of nominal labels. Convex objective functions based on misclassification error minimization (or approximation) are not that different either. Least-squares or logistic regression methods set up convex objective functions with nominal labels converted to binary outputs [27, 28]. When extensions to multi-class are sought, the binary labels are extended to one of K encoding, with K being the number of classes.

Support vector machines (SVMs) were inherently designed for two-class discrimination, and all formulations of multi-class SVMs extend this oppositional framework using one-versus-one or one-versus-all schemes. Below, we begin by describing the different approaches to the multi-class problem. This is not meant to be exhaustive but provides an overview of some popular methods and approaches researched in classification and dimensionality reduction. Folley and Sammon [29, 30] studied the two class problems and feature selection and focused on criteria with the greatest potential to discriminate. Feature selection aims to find a set of features with the best discrimination properties. To identify the best feature vectors, they chose the generalized Fisher optimality criterion proposed by [31]. The selected directions maximize the Fisher criterion, which has attractive discrimination properties. Principal components analysis (PCA) permits the reduction of dimensions of high dimensional data without losing significant information [16, 20, 32]. Principal components identify patterns or significant features without considering discriminative considerations [33]. Supervised PCA (SPCA), derived from PCA, is a method for obtaining useful sub-spaces when the labels are considered. This technique was first described in [34] under the title "supervised clustering." The idea behind SPCA is to perform selective dimensionality reduction using carefully chosen subsets of labeled samples. This is used to build a prediction model [35]. Unfortunately, despite the superficial similarity to our work, SPCA does not carefully reformulate supervised dimensionality reduction as an optimization problem (as we do) and develop an algorithm that projects vectors into the space of categories- the present work's goal. While we have addressed the most popular techniques in dimensionality reduction and multi-class classification, this is not an exhaustive study of the literature. We focus primarily on discriminative dimensionality reduction methods that improve multi-class classification performance. The closest we have seen in relation to our work on category spaces is the work in [36, 37]. They mention the importance and usefulness of modeling categories as vector spaces for document retrieval and explain how unrelated items should have an orthogonal relationship. This is to say that they should have no features in common. The structured SVM in [38] is another effort at going beyond nominal classes. Here, classes can have internal structures in the form of strings, trees, etc. However, explicitly modeling classes as vector spaces is not carried out.

From the above, the modest goal of the present work should be clear. We seek to project the input feature vectors to a category space—, a subspace formed by category basis vectors. The multi-class FLD falls short of this goal since the number of projected dimensions is one less than the number of classes. The multi-class (and more recently multi-label) SVM [39] literature is fragmented due to a lack of agreement regarding the core issue of multi-class discrimination. The varieties of supervised PCA do not begin by clearly formulating a criterion for category space projection. Variants such as CCA [40, 41], PLS [42, 43], and structured SVM's [38] while attempting to add structure to the categories do not go as far as the present work in attempting to fit a category subspace. Kernel variants of the above also do not touch the basic issue addressed in the present work. Nonlinear (and manifold learning-based) dimensionality reduction techniques [8, 11, 13, 44, 45] are unsupervised and therefore do not qualify.

3. Dimensionality reduction using a category space formulation

3.1 Maximizing the square of the inner product

The principal goal is a new form of supervised dimensionality reduction. Specifically, when we seek to marry principal component analysis with supervised learning, the simplest synthesis is category space dimensionality reduction with orthogonal class vectors. Assume the existence of a feature space with each feature vector $x_i \in \mathbf{R}^D$. We aim to perform supervised dimensionality reduction by reducing the number of feature dimensions from D to K where $K \leq D$. Here K is the number of classes, and the first simplifying assumption made in this work is that we will represent the category space using K *orthonormal* basis vectors $\{w_k\}$ together with an *origin* $x_0 \in \mathbf{R}^D$. The second assumption we make is that each feature vector x_i should have a large magnitude inner product with its assigned class. From the orthonormality constraint above, this automatically implies a small magnitude inner product with all other weight vectors. A *candidate objective function* and constraints following the above considerations are

$$E(W) = -\frac{1}{2}\sum_{k=1}^{K}\sum_{i_k \in C_k}\left[w_k^T\left(x_{i_k} - x_0\right)\right]^2 \tag{1}$$

and

$$w_k^T w_l = \begin{cases} 1, & k = l \\ 0, & k \neq l \end{cases} \tag{2}$$

respectively. In Eq. (1), $W = [w_1, w_2, \ldots, w_K]$. Note that we have referred to this as a candidate objective function for two reasons. First, the origin x_0 is still unspecified, and obviously, we cannot minimize Eq. (1) w.r.t. x_0 as the minimum value is not bounded from below. Second, it is unclear why we cannot use the inner product's absolute value or other symmetric functions. Both these issues are addressed later in this work. We currently resolve the origin issue by setting x_0 to the centroid of all the feature vectors (with this choice getting a principled justification below).

The objective function in Eq. (1) is the negative of a quadratic function. Since the function $-x^2$ is concave, it admits a Legendre transform-based majorization [46] using the tangent of the function. That is, we propose to replace objective functions of the form $-\frac{1}{2}x^2$ with $\min_y -xy + \frac{1}{2}y^2$ which can quickly be checked to be valid for an unconstrained auxiliary variable y. Note that this transformation yields a linear objective function w.r.t. x, which is to be expected from the geometric interpretation of a tangent.

Consider the following Legendre transformation of the objective function in Eq. (1). The new objective function is

$$E_{\text{quad}}(W, Z) = \sum_{k=1}^{K} \sum_{i_k \in C_k} \left[z_{ki_k} \left(-w_k^T x_{i_k} + w_k^T x_0 \right) + \frac{1}{2} z_{ki_k}^2 \right] \tag{3}$$

where $Z = \{ z_{ki_k} | k \in \{1, \dots, K\}, i_k \in \{1, \dots, |C_k|\} \}$. Consider this an objective function over x_0 as well. We require additional constraints to avoid minima at negative infinity (i.e., for the objective function to be *coercive*). One such constraint (and perhaps not the only one) is of the form $\sum_{i_k \in C_k} z_{ki_k} = 0, \forall k$. When this constraint is imposed, we obtain a new objective function,

$$E_{\text{quad}}(W, Z) = \sum_{k=1}^{K} \sum_{i_k \in C_k} \left[-z_{ki_k} w_k^T x_{i_k} + \frac{1}{2} z_{ki_k}^2 \right] \tag{4}$$

to be minimized subject to the constraints

$$\sum_{i_k \in C_k} z_{ki_k} = 0, \forall k \tag{5}$$

which are in addition to the orthonormal constraints in Eq. (2). This objective function yields a Z, which removes the class-specific centroid of C_k for all classes.

3.2 Maximizing the absolute value of the inner product

We have justified our choice of centroid removal mentioned above indirectly obtained via constraints imposed on Legendre transform auxiliary variables. The above objective function can be suitably modified using different forms (absolute inner product, etc.). To see this, consider the following objective function, which minimizes the negative of the magnitude of the inner product:

$$E(W) = -\sum_{k=1}^{K} \sum_{i_k \in C_k} | w_k^T (x_{i_k} - x_0) |. \tag{6}$$

Since $-|x|$ is also a concave function, it too can be majorized. Consider first replacing the non-differentiable objective function $-|x|$ with $-\sqrt{x^2 + \epsilon}$ (also concave) where ϵ can be chosen to be a suitably small value. Now consider replacing $-\sqrt{x^2 + \epsilon}$ with $\min_y -xy - \epsilon\sqrt{1 - y^2}$ which can again quickly checked to be valid for a constrained auxiliary variable $y \in [-1, 1]$. The constraint is somewhat less relevant since the minimum w.r.t. y occurs at $y = \frac{x}{\sqrt{x^2+\epsilon^2}}$ which lies within the constraint interval. Note that this transformation yields a linear objective function w.r.t. x. As before, we introduce a new objective function

$$E_{\text{abs}}(W, Z) = \sum_{k=1}^{K} \sum_{i_k \in C_k} \left[-z_{ki_k} w_k^T x_{i_k} - \epsilon \sqrt{1 - z_{ki_k}^2} \right] \tag{7}$$

to be minimized subject to the constraints $\sum_{i_k \in C_k} z_{ki_k} = 0, \forall k$ and $z_{ki_k} \in [-1, 1]$ which are the same as in Eq. (5) in addition to the orthonormal constraints in Eq. (2).

3.3 Extension to RKHS kernels

The generalization to RKHS kernels is surprisingly straightforward. First, we follow standard kernel PCA and write the weight vector in terms of the RKHS projected patterns $\phi(x_l)$ to get

$$w_k = \sum_{i=1}^{N} \alpha_{ki} \phi(x_i). \tag{8}$$

Note that the expansion of the weight vector is in overall patterns rather than just the class-specific ones. This assumes that the weight vector for each class lives in the subspace (potentially infinite dimensional) spanned by the RKHS projected patterns —the same assumption as in standard kernel PCA. The orthogonality constraint between weight vectors becomes

$$\begin{aligned} \langle w_k, w_l \rangle &= \left\langle \sum_{i=1}^{N} \alpha_{ki} \phi(x_i), \sum_{i=1}^{N} \alpha_{li} \phi(x_i) \right\rangle \\ &= \sum_{i=1}^{N} \sum_{j=1}^{N} \alpha_{ki} \alpha_{kj} \langle \phi(x_i), \phi(x_j) \rangle \\ &= \sum_{i=1}^{N} \sum_{j=1}^{N} \alpha_{ki} \alpha_{kj} K(x_i, x_j) \end{aligned} \tag{9}$$

which is equal to one if $k = l$ and zero otherwise. In matrix form, the orthonormality constraints become

$$AGA^T = I_K \tag{10}$$

where $[A]_{kl} \equiv \alpha_{ki}$ and $[G]_{ij} = K(x_i, x_j)$ is the well-known Gram matrix of pairwise RKHS inner products between the patterns.

The corresponding squared inner product and the absolute value of the inner product objective functions are

$$E_{\text{Kquad}}(A, Z) = \sum_{k=1}^{K} \sum_{i_k \in C_k} \left[-\sum_{j=1}^{N} z_{ki_k} \alpha_{kj} K(x_j, x_{i_k}) + \frac{1}{2} z_{ki_k}^2 \right] \tag{11}$$

and

$$E_{\mathrm{Kabs}}(A,Z) = \sum_{k=1}^{K} \sum_{i_k \in C_k} \left[-\sum_{j=1}^{N} z_{ki_k} \alpha_{kj} K\left(x_j, x_{i_k}\right) - \epsilon \sqrt{1 - z_{ki_k}^2} \right] \tag{12}$$

respectively. These have to be minimized w.r.t. the orthonormal constraints in Eq. (10) and the origin constraints in Eq. (5). Note that the objective functions are identical w.r.t. the matrix A. The parameter ϵ can be set to a very small but positive value.

4. An algorithm for supervised dimensionality reduction

We now return to the objective functions and constraints in Eqs. (4) and (7) prior to tackling the corresponding kernel versions in Eqs. (11) and (12) respectively. It turns out that the approach for minimizing the former can be readily generalized to the latter, with the former being easier to analyze. Note that the objective functions in Eqs. (4) and (7) are identical w.r.t. W. Consequently, we dispense with the optimization problems w.r.t. Z, which are straightforward and focus on the optimization problem w.r.t. W.

4.1 Weight matrix estimation with orthogonality constraints

The objective function and constraints on W can be written as

$$E_{\mathrm{equiv}}(W) = -\sum_{k=1}^{K} \sum_{i_k \in C_k} z_{ki_k} w_k^T x_{i_k} \tag{13}$$

and

$$w_k^T w_l = \begin{cases} 1, & k = l \\ 0, & k \neq l \end{cases}. \tag{14}$$

Note that the set Z is not included in this objective function despite its presence in the larger objective functions of Eqs. (4) and (7). The orthonormal constraints can be expressed using a Lagrange parameter matrix to obtain the following Lagrangian:

$$L(W,\Lambda) = -\sum_{k=1}^{K} \sum_{i_k \in C_k} z_{ki_k} w_k^T x_{i_k} + \mathrm{trace}\left\{\Lambda\left(W^T W - I_K\right)\right\}. \tag{15}$$

Setting the gradient of L w.r.t. W to zero, we obtain

$$\nabla_W L(W,\Lambda) = -Y + W\left(\Lambda + \Lambda^T\right) = 0 \tag{16}$$

where the matrix Y of size $D \times K$ is defined as

$$Y \equiv \left[\sum_{i_1 \in C_1} z_{1i_1} x_{i_1}, \ldots, \sum_{i_k \in C_k} z_{ki_k} x_{i_k} \right]. \tag{17}$$

Using the constraint $W^TW = I_K$, we get

$$\left(\Lambda + \Lambda^T\right) = W^TY. \tag{18}$$

Since $\left(\Lambda + \Lambda^T\right)$ is symmetric, this immediately implies that W^TY is symmetric. From Eq. (16), we also get

$$\left(\Lambda + \Lambda^T\right)W^TW\left(\Lambda + \Lambda^T\right) = \left(\Lambda + \Lambda^T\right)^2 = Y^TY. \tag{19}$$

Expanding Y using its singular value decomposition (SVD) as $Y = U\Sigma V^T$, the above relations can be simplified to

$$Y = U\Sigma V^T = UV^T\left(V\Sigma V^T\right) = W\left(\Lambda + \Lambda^T\right) \tag{20}$$

giving

$$\left(\Lambda + \Lambda^T\right) = V\Sigma V^T \tag{21}$$

and

$$W = UV^T. \tag{22}$$

We have shown that the optimal solution for W is the polar decomposition of Y, namely $W = UV^T$. Since Z has been held fixed during the estimation of W, in the subsequent step, we can hold W fixed and solve for Z and repeat. We obtain an alternating algorithm that iterates between estimating W and Z until a convergence criterion is met.

4.2 Estimation of the auxiliary variable Z

The objective function and constraints on Z depend on whether we use objective functions based on the square or absolute value of the inner product. We separately consider the two cases. The inner product squared effective objective function

$$E_{\text{quadeff}}(Z) = \sum_{k=1}^{K}\sum_{i_k \in C_k}\left[-z_{ki_k}w_k^Tx_{i_k} + \frac{1}{2}z_{ki_k}^2\right] \tag{23}$$

is minimized w.r.t. Z subject to the constraints $\sum_{i_k \in C_k} z_{ki_k} = 0, \forall k$. The straightforward solution obtained via standard minimization is

$$\begin{aligned} z_{ki_k} &= w_k^Tx_{i_k} - \frac{1}{|C_k|}\sum\nolimits_{i_k \in C_k} w_k^Tx_{i_k} \\ &= w_k^T\left(x_{i_k} - \frac{1}{|C_k|}\sum\nolimits_{i_k \in C_k} x_{i_k}\right). \end{aligned} \tag{24}$$

The absolute value effective objective function

$$E_{\text{abseff}}(Z) = \sum_{k=1}^{K}\sum_{i_k \in C_k}\left[-z_{ki_k}w_k^Tx_{i_k} - \epsilon\sqrt{1 - z_{ki_k}^2}\right] \tag{25}$$

is also minimized w.r.t. Z subject to the constraints $\sum_{i_k \in C_k} z_{ki_k} = 0, \forall k$. A heuristic solution obtained (eschewing standard minimization) is

$$z_{ki_k} = \frac{w_k^T x_{i_k}}{\sqrt{\left(w_k^T x_{i_k}\right)^2 + \epsilon^2}} - \frac{1}{|C_k|} \sum_{i_k \in C_k} \frac{w_k^T x_{i_k}}{\sqrt{\left(w_k^T x_{i_k}\right)^2 + \epsilon^2}} \tag{26}$$

which has to be checked to be valid. The heuristic solution acts as an initial condition for constraint satisfaction (which can be efficiently obtained via 1D line minimization). The first order Karush-Kuhn-Tucker (KKT) conditions obtained from the Lagrangian

$$L_{\text{abseff}}(Z, M) = \sum_{k=1}^{K} \sum_{i_k \in C_k} \left[-z_{ki_k} w_k^T x_{i_k} - \epsilon \sqrt{1 - z_{ki_k}^2} \right] - \sum_{k=1}^{K} \mu_k \sum_{i_k \in C_k} z_{ki_k} \tag{27}$$

are

$$-w_k^T x_{i_k} + \epsilon \frac{z_{ki_k}}{\sqrt{1 - z_{ki_k}^2}} - \mu_k = 0, \forall k \tag{28}$$

from which we obtain

$$z_{ki_k} = \frac{w_k^T x_{i_k} + \mu_k}{\sqrt{\left(w_k^T x_{i_k} + \mu_k\right)^2 + \epsilon^2}}. \tag{29}$$

We see that the constraint $z_{ki_k} \in [-1, 1]$ is also satisfied. For each category C_k, there exists a solution to the Lagrange parameter μ_k such that $\sum_{i_k} z_{ki_k} = 0$. This can be obtained via any efficient 1D search procedure like golden section [47].

4.3 Extension to the kernel setting

The objective function and constraints on the weight matrix A in the kernel setting are

$$E_{\text{Kequiv}}(A) = -\sum_{k=1}^{K} \sum_{i_k \in C_k} \sum_{j=1}^{N} z_{ki_k} \alpha_{kj} K\left(x_j, x_{i_k}\right) \tag{30}$$

with the constraints

$$AGA^T = I_K \tag{31}$$

where $[A]_{ki} = \alpha_{ki}$ and $[G]_{ij} = K\left(x_i, x_j\right)$ is the $N \times N$ kernel Gram matrix. The constraints can be expressed using a Lagrange parameter matrix to obtain the following Lagrangian,

$$L_{\text{ker}}(A, \Lambda) = -\sum_{k=1}^{K} \sum_{i_k \in C_k} \sum_{j=1}^{N} z_{ki_k} \alpha_{kj} K\left(x_j, x_{i_k}\right) + \text{trace}\left\{\Lambda_{\text{ker}}\left(AGA^T - I_K\right)\right\}. \tag{32}$$

Setting the gradient of L_{ker} w.r.t. A to zero, we obtain

$$-Y_{\text{ker}} + \left(\Lambda_{\text{ker}} + \Lambda_{\text{ker}}^T\right)AG = 0 \tag{33}$$

where the matrix Y_{ker} of size $K \times N$ is defined as

$$[Y_{\text{ker}}]_{kj} \equiv \sum_{i_k \in C_k} z_{ki_k} K\left(x_j, x_{i_k}\right). \tag{34}$$

Using the constraint $AGA^T = I_K$, we obtain

$$\left(\Lambda_{\text{ker}} + \Lambda_{\text{ker}}^T\right)AGA^T\left(\Lambda_{\text{ker}} + \Lambda_{\text{ker}}^T\right) = \left(\Lambda_{\text{ker}} + \Lambda_{\text{ker}}^T\right)^2 = Y_{\text{ker}}G^{-1}Y_{\text{ker}}^T. \tag{35}$$

Expanding $Y_{\text{ker}}G^{-\frac{1}{2}}$ using its singular value decomposition as $Y_{\text{ker}}G^{-\frac{1}{2}} = U_{\text{ker}}S_{\text{ker}}V_{\text{ker}}^T$, the above relations can be simplified to

$$\left(\Lambda_{\text{ker}} + \Lambda_{\text{ker}}^T\right) = U_{\text{ker}}S_{\text{ker}}U_{\text{ker}}^T \tag{36}$$

and

$$AG^{\frac{1}{2}} = U_{\text{ker}}V_{\text{ker}}^T \Rightarrow A = U_{\text{ker}}V_{\text{ker}}^T G^{-\frac{1}{2}}. \tag{37}$$

We have shown that the optimal solution for A is related to the polar decomposition of $Y_{\text{ker}}G^{-\frac{1}{2}}$, namely $A = U_{\text{ker}}V_{\text{ker}}^T G^{-\frac{1}{2}}$. Since Z has been held fixed during the estimation of A, in the subsequent step, we can hold A fixed and solve for Z and repeat. We thereby obtain an alternating algorithm that iterates between estimating A and Z until a convergence criterion is met. This is analogous to the non-kernel version above.

The solutions for Z in this setting are very straightforward to obtain. We eschew the derivation and merely state that

$$\begin{aligned} z_{ki_k} &= \sum_{j=1}^N \alpha_{kj} K\left(x_j, x_{i_k}\right) - \frac{1}{|C_k|}\sum_{i_k \in C_k}\sum_{j=1}^N \alpha_{kj}K\left(x_j, x_{i_k}\right) \\ &= \sum_{j=1}^N \alpha_{kj}\left(K\left(x_j, x_{i_k}\right) - \frac{1}{|C_k|}\sum_{i_k \in C_k} K\left(x_j, x_{i_k}\right)\right) \end{aligned} \tag{38}$$

for the squared inner product kernel objective and

$$z_{ki_k} = \frac{\sum_{j=1}^N \alpha_{kj}K\left(x_j, x_{i_k}\right)}{\sqrt{\left(\sum_{j=1}^N \alpha_{kj}K\left(x_j, x_{i_k}\right)\right)^2 + \epsilon^2}} - \frac{1}{|C_k|}\sum_{i_k \in C_k} \frac{\sum_{j=1}^N \alpha_{kj}K\left(x_j, x_{i_k}\right)}{\sqrt{\left(\sum_{j=1}^N \alpha_{kj}K\left(x_j, x_{i_k}\right)\right)^2 + \epsilon^2}} \tag{39}$$

for the absolute valued kernel objective. This heuristic solution acts as an initial condition for constraint satisfaction (which can be efficiently obtained via 1D line minimization). Following the line of Eqs. (27)–(29) above, the solution can be written as

$$z_{ki_k} = \frac{\sum_{j=1}^N \alpha_{kj}K\left(x_j, x_{i_k}\right) + \mu_k}{\sqrt{\left(\sum_{j=1}^N \alpha_{kj}K\left(x_j, x_{i_k}\right) + \mu_k\right)^2 + \epsilon^2}}. \tag{40}$$

For each category C_k, as before, there exists a solution to the Lagrange parameter μ_k such that $\sum_{i_k} z_{ki_k} = 0$. Once again, this can be obtained via any efficient 1D search procedure like golden section [47].

4.4 Analysis

4.4.1 Euclidean setting

The simplest objective function in the above sequence, analyzed in the literature, is based on the squared inner product. Below, we summarize this work by closely following the treatment in [2, 48]. First, to bring our work in sync with the literature, we eliminate the auxiliary variable Z from the squared inner product objective function (treated as a function of both W and Z here):

$$E_{\text{quadeff}}(W,Z) = \sum_{k=1}^{K} \sum_{i_k \in C_k} \left[-z_{ki_k} w_k^T x_{i_k} + \frac{1}{2} z_{ki_k}^2 \right]. \tag{41}$$

Setting $z_{ki_k} = w_k^T \left(x_{i_k} - \frac{1}{|C_k|} \sum_{i_k \in C_k} x_{i_k} \right)$ which is the optimum solution for Z, we get

$$E_{\text{quad}}(W) = -\frac{1}{2} \sum_{k=1}^{K} w_k^T R_k w_k \equiv -\frac{1}{2} \sum_{k=1}^{K} \sum_{i_k \in C_k} \left[w_k^T \left(x_{i_k} - \frac{1}{|C_k|} \sum_{i \in C_k} x_i \right) \right]^2 \tag{42}$$

where R_k is the class-specific covariance matrix:

$$R_k \equiv \sum_{i_k \in C_k} \left(x_{i_k} - \frac{1}{|C_k|} \sum_{i \in C_k} x_i \right) \left(x_{i_k} - \frac{1}{|C_k|} \sum_{i \in C_k} x_i \right)^T. \tag{43}$$

We seek to minimize Eq. (42) w.r.t. W under the orthonormality constraints $W^T W = I_K$.

A set of K orthonormal vectors $\{ w_k \in \mathbf{R}^D, k \in \{1, \ldots, K\} \}$ in a D-dimensional Euclidean space is a point on the well-known Stiefel manifold, denoted here by $M_{D,K}$ with $K \le D$. The problem in Eq. (42) is equivalent to the maximization of the sum of heterogeneous quadratic functions on a Stiefel manifold. The functions are heterogeneous in our case since the class-specific covariance matrices R_k are not identical in general. The Lagrangian corresponding to this problem (with Z removed via direct minimization) is

$$L_{\text{quad}}(W, \Lambda) = -\frac{1}{2} \sum_{k=1}^{K} w_k^T R_k w_k + \text{trace}\left[\Lambda^T \left(W^T W - I_K \right) \right]. \tag{44}$$

Setting the gradient of the above Lagrangian w.r.t. W to zero, we obtain

$$[R_1 w_1, R_2 w_2, \ldots, R_K w_K] = W\left(\Lambda + \Lambda^T\right). \tag{45}$$

Noting that $\Lambda + \Lambda^T$ is symmetric and using the Stiefel orthonormality constraint $W^T W = I_K$, we get

$$(\Lambda + \Lambda^T) = W^T[R_1 w_1, R_2 w_2, \ldots, R_K w_K]. \tag{46}$$

The above can be considerably simplified. First we introduce a new vector $\mathbf{w} \in M_{D,K}$ defined as $\mathbf{w} \equiv \left[w_1^T, w_2^T, \ldots, w_K^T\right]^T$ and then rewrite Eq. (45) in vector form to get

$$R\mathbf{w} = S(\mathbf{w})\mathbf{w} \tag{47}$$

where

$$R \equiv \begin{bmatrix} R_1 & 0_K & \cdots & 0_K \\ 0_K & R_2 & \cdots & 0_K \\ 0_K & \cdots & \ddots & 0_K \\ 0_K & \cdots & \cdots & R_K \end{bmatrix} \tag{48}$$

is a $KD \times KD$ matrix and

$$S(\mathbf{w}) \equiv \begin{bmatrix} w_1^T R_1 w_1 I_K & \cdots & \frac{1}{2}\left(w_1^T R_1 w_K + w_K^T R_K w_1\right) I_K \\ \frac{1}{2}\left(w_1^T R_1 w_2 + w_2^T R_2 w_1\right) I_K & \cdots & \frac{1}{2}\left(w_2^T R_2 w_K + w_K^T R_K w_2\right) I_K \\ \vdots & \ddots & \vdots \\ \frac{1}{2}\left(w_1^T R_1 w_K + w_K^T R_K w_1\right) I_K & \cdots & w_K^T R_K w_K I_K \end{bmatrix} \tag{49}$$

a $KD \times KD$ *symmetric* matrix. The reason $S(\mathbf{w})$ can be made symmetric is because it is closely related to the solution to $(\Lambda + \Lambda)^T$, which has to be symmetric. The first and second order necessary conditions for a vector $\mathbf{w}_0 \in M_{D,K}$ to be a local minimum (feasible point) for the problem in Eq. (42) are as follows:

$$R\mathbf{w}_0 = S(\mathbf{w}_0)\mathbf{w}_0 \tag{50}$$

and

$$(R - S(\mathbf{w}_0))|_{TM(\mathbf{w}_0)} \tag{51}$$

is negative semi-definite. In Eq. (51), $TM(\mathbf{w}_0)$ is the tangent space of the Stiefel manifold at $\mathbf{w}_0$. In a *tour de force* proof, Rapcsák further shows in [2] that if the matrix $(R - S(\mathbf{w}_0))$ is negative semi-definite, then a feasible point $\mathbf{w}_0$ is a *global minimum*. This is an important result since it adds a sufficient condition for a global minimum for the problem of minimizing a heterogeneous sum of quadratic forms on a Stiefel manifold.[1]

---

[1] Note that this problem is fundamentally different from and cannot be reduced to the minimization of trace (AW^TBW) subject to $W^TW = I_K$ which has a closed form solution.

4.4.2 The RKHS setting

We can readily extend the above analysis to the kernel version of the squared inner product. The complete objective function w.r.t. both the coefficients A and the auxiliary variable Z is

$$E_{\text{Kequiv}}(A,Z) = \sum_{k=1}^{K} \sum_{i_k \in C_k} \left[-\sum_{j=1}^{N} z_{ki_k} \alpha_{kj} K\left(x_j, x_{i_k}\right) + \frac{1}{2} z_{ki_k}^2 \right]. \tag{52}$$

Setting $z_{ki_k} = \sum_{j=1}^{N} \alpha_{kj} K\left(x_j, x_{i_k}\right)$ which is the optimum solution for Z, we get

$$\begin{aligned} E_{\text{Kquad}}(A) &= -\frac{1}{2} \sum_{k=1}^{K} \sum_{i_k \in C_k} \left[\sum_{j=1}^{N} \alpha_{kj} \left(K\left(x_j, x_{i_k}\right) - \frac{1}{|C_k|} \sum_{i_k \in C_k} K\left(x_j, x_{i_k}\right) \right) \right]^2 \\ &= -\frac{1}{2} \sum_{k=1}^{K} \boldsymbol{\alpha}_k^T G_k \boldsymbol{\alpha}_k \end{aligned} \tag{53}$$

where $[\boldsymbol{\alpha}_k]_j = \alpha_{kj}, A = [\boldsymbol{\alpha}_1, \boldsymbol{\alpha}_2, \ldots, \boldsymbol{\alpha}_K]^T$ (and)

$$[G_k]_{jm} \equiv \sum_{i_k \in C_k} \left(K\left(x_j, x_{i_k}\right) - \frac{1}{|C_k|} \sum_{i \in C_k} K\left(x_j, x_i\right) \right) \cdot \left(K\left(x_m, x_{i_k}\right) - \frac{1}{|C_k|} \sum_{i \in C_k} K\left(x_m, x_i\right) \right) \tag{54}$$

The constraints on A can be written as

$$AGA^T = I_K \Rightarrow \left(G^{\frac{1}{2}} A^T\right)^T \left(G^{\frac{1}{2}} A^T\right) = I_K. \tag{55}$$

Introducing a new variable $B = G^{\frac{1}{2}} A^T$, we may rewrite the kernel objective function and constraints as

$$E_{\text{Kquadnew}}(B) = -\frac{1}{2} \sum_{k=1}^{K} \boldsymbol{\beta}_k^T H \boldsymbol{\beta}_k \equiv -\frac{1}{2} \sum_{k=1}^{K} \boldsymbol{\beta}_k^T G^{-\frac{1}{2}} G_k G^{-\frac{1}{2}} \boldsymbol{\beta}_k \tag{56}$$

(where $B \equiv [\boldsymbol{\beta}_1, \boldsymbol{\beta}_2, \ldots, \boldsymbol{\beta}_K]$) and

$$B^T B = I_K \tag{57}$$

respectively. This is now in the same form as the objective function and constraints in Section 4.4.1, and therefore, the Rapcsák analysis of that section can be directly applied here. The above change of variables is predicated on the positive definiteness of G. If this is invalid, principal component analysis has to be applied to G, resulting in a positive definite matrix in a reduced space, after which the above approach can be applied.

In addition to providing necessary conditions for global minima, the authors in [4] developed an iterative procedure as a method for a solution. We have adapted this to suit our purposes. A block coordinate descent algorithm which successively updates W and Z is presented in Algorithm 1.

Algorithm 1. Iterative process for minimization of the sum of squares of inner products objective function.

- **Input**: A set of labeled patterns $\{x_{i_k}\}_{i_k=1}^{|C_k|}, \forall k \in \{1, \dots, K\}$.
- **Initialize**:
  - Convergence threshold δ.
  - Arbitrary orthonormal system $W^{(0)}$.
- **Repeat**
  - Calculate the sequence $\{W^{(1)}, W^{(2)}, \dots, W^{(m)}\}$. Assume $W^{(m)}$ is constructed for $m = 0,1,2, \dots$
  - Update the auxiliary variable $Z^{(m+1)}$ satisfying $\sum_{i_k \in C_k} z_{ki_k} = 0$:
  - $* z_{ki_k}^{(m+1)} = \left(w^{(m)}\right)_k^T x_{i_k} - \frac{1}{|C_k|} \sum_{i_k \in C_k} \left(w^{(m)}\right)_k^T x_{i_k}$.
  - Perform the SVD decomposition on $\left[\sum_{i_1 \in C_1} z_{1i_1}^{(m+1)} x_{i_1}, \dots, \sum_{i_k \in C_k} z_{ki_k}^{(m+1)} x_{i_k}\right]$ to get $U^{(m+1)} S^{(m+1)} \left(V^{(m+1)}\right)^T$ where $S^{(m+1)}$ is $K \times K$.
  - $W^{(m+1)} = U^{(m+1)} \left(V^{(m+1)}\right)^T$, the polar decomposition.
- **Loop until** $\left\|W^{(m+1)} - W^{(m)}\right\|_{\mathcal{F}} \leq \delta$.
- **Output**: W

5. Experimental results

5.1 Quantitative results for linear and kernel dimensionality reduction

In practice, dimensionality reduction is used in conjunction with a classification algorithm. By definition, the purpose of dimensionality reduction, as it relates to classification, is to reduce the complexity of the data while retaining discriminating information. Thus, we utilize a popular classification algorithm to analyze the performance of our proposed dimensionality reduction technique. In this section, we report the results of several experiments with dimensionality reduction combined with SVM classification. In the multi-class setting, we compare against other state-of-the-art algorithms that perform dimensionality reduction and then evaluate the performance using the multi-class one-vs-all linear SVM scheme. The classification technique uses the traditional training and testing phases, outputting the class it considers the best prediction for a given test sample. We measure the accuracy of these predictions averaged over all test sets. In **Table 1**, we demonstrate the effectiveness of both the sum of quadratic and absolute value functions, denoted as category quadratic space (CQS) and category absolute value space (CAS), respectively. Then, we benchmark their overall classification accuracy against several classical dimensionality reduction techniques, namely, least squares linear discriminant analysis (LS-LDA) [27], Fisher linear discriminant (MC-FLD) [19], principal component analysis (PCA) [33] and their multi-class and kernel counterparts (when applicable). In each experiment, we chose two-thirds of the data for training, and the remaining third of the samples were used for testing. The results are shown in **Table 1**.

Databases: To illustrate the performance of the methods proposed in Section 3, we conducted experiments using different publicly available data sets taken from the UCI machine learning data repository [49]. We have chosen a variety of data sets that vary in terms of class cardinality (K), samples (N), and number of features (D) to demonstrate the versatility of our approach. For a direct comparison of results, we chose the same data sets: Vehicle, Wine, Iris, Seeds, Thyroid, Satellite, Segmentation, and Vertebral Silhouettes recognition databases. More details about the individual sets are available at the respective repository sites.

We divide the results into the linear and kernel groups (as is normal practice). **Table 1** shows the results for linear dimensionality reduction with SVM linear classification. All dimensionality reduction algorithms were implemented and configured with a linear SVM classifier for optimal classification results (via cross-validation). It can be seen that the category space projection scheme consistently provides a good projection for standard classification algorithms to be executed. Several of the data sets comprise only three classes, and it can be seen that the proposed method is competitive in performance and, in some instances, performs slightly better.

Also, for comparison, **Table 2** reports the performance of the proposed kernel formulations followed by a linear SVM classifier. These proposed methods also achieve accuracy rates similar to their kernel counterparts with angle classification shown in **Table 3**.

The iterative approach in Algorithm (1) was applied to obtain an optimal orthonormal basis W (which is $D \times K$) for the category space, where D dimensional input

Name (#Classes)	CQS	CAS	LS-LDA	PCA	MC-FLD
Vehicle (4)	53.91	53.05	76.56	55.36	76.82
Wine (3)	96.07	96.82	95.51	77.19	97.28
Iris (3)	97.55	96.88	96.11	96.77	96.77
Seeds (3)	90.39	90.79	95.15	92.53	95.79
thyroid (3)	94.02	94.08	94.02	92.57	93.92
Satellite (6)	85.30	85.20	86.38	85.45	86.52
Segmentation (7)	93.14	93.44	94.62	94.40	94.43
Vertebral (3)	84.13	82.79	81.45	80.05	81.18

Table 1.
Linear dimensionality reduction w/SVM classification.

Name (# Classes)	K-CQS	K-CAS	K-PCA	K-MC-FLD
Vehicle (4)	40.27	40.92	44.81	74.35
Wine (3)	92.95	95.63	95.95	96.88
Iris (3)	95.55	93.33	95.55	94.44
Seeds (3)	90.21	90.47	91.53	93.65
thyroid (3)	41.97	40.24	43.08	72.34
Satellite (6)	81.54	86.23	89.69	90.61
Segmentation (7)	72.96	77.24	83.01	92.43
Vertebral (3)	70.96	69.53	70.96	82.25

Table 2.
Kernel dimensionality reduction w/SVM classification.

Name (# Classes)	K-CQS-A	K-CAS-A
Vehicle (4)	67.96	68.24
Wine (3)	95.32	95.32
Iris (3)	95.55	95.18
Seeds (3)	91.79	91.79
thyroid (3)	67.90	66.79
Satellite (6)	83.33	76.29
Segmentation (7)	50.21	48.94
Vertebral (3)	77.59	77.77

Table 3.
Kernel dimensionality reduction w/angle classification.

patterns can be projected to the smaller K dimensional category space if $D > K$. We start with a set of N labeled, input vectors $x_i \in \mathbf{R}^D$ drawn randomly from multiple classes $C_k, k \in \{1, \dots, K\}$. The optimization technique searches over Steifel manifold elements, as explained above. The algorithm is terminated when the Frobenius norm difference between iterations, $\|W^{(m-1)} - W^{(m)}\|_F \leq \delta$ (with $\delta = 10^{-8}$). Once we have determined the optimal W, the patterns are mapped to the category space by the transformation $y_i = W^T x_i$ to obtain the corresponding set of N samples $y_i \in \mathbf{R}^K$, where K is the reduced dimensional space.

The results above show that our proposed methods lead to classification rates that can be compared to classical approaches. However, the main focus of this work is to provide an algorithm that retains important classification information while introducing a geometry (category vector subspace) with attractive semantic and visualization properties. The results suggest that our classification results are competitive with other techniques while learning a category space.

5.2 Visualization of kernel dimensionality reduction

Another valuable aspect of this research can be seen in the kernel formulation, which demonstrates the warping of the projected patterns toward their respective category axes. This suggests a geometric approach to classification, i.e., we could consider the angle of deviation of a test set pattern from each category axis to measure class membership. Within the category space, a base category is represented by the bases (axes) that define the category space. Class membership is, therefore, inversely proportional to the angle between the pattern and the respective category axis. **Figures 1** through **3** illustrate the warped space for various three class problems, for a variation in the width parameter (σ) of a Gaussian radial basis function kernel in the range $\sigma = [0.1, 0.8]$. Note the improved visualization semantics of the category space approach when compared to the other dimensionality reduction techniques.

6. Conclusions

In this work, we presented a new approach to supervised dimensionality reduction—that attempts to learn orthogonal category axes during training. The

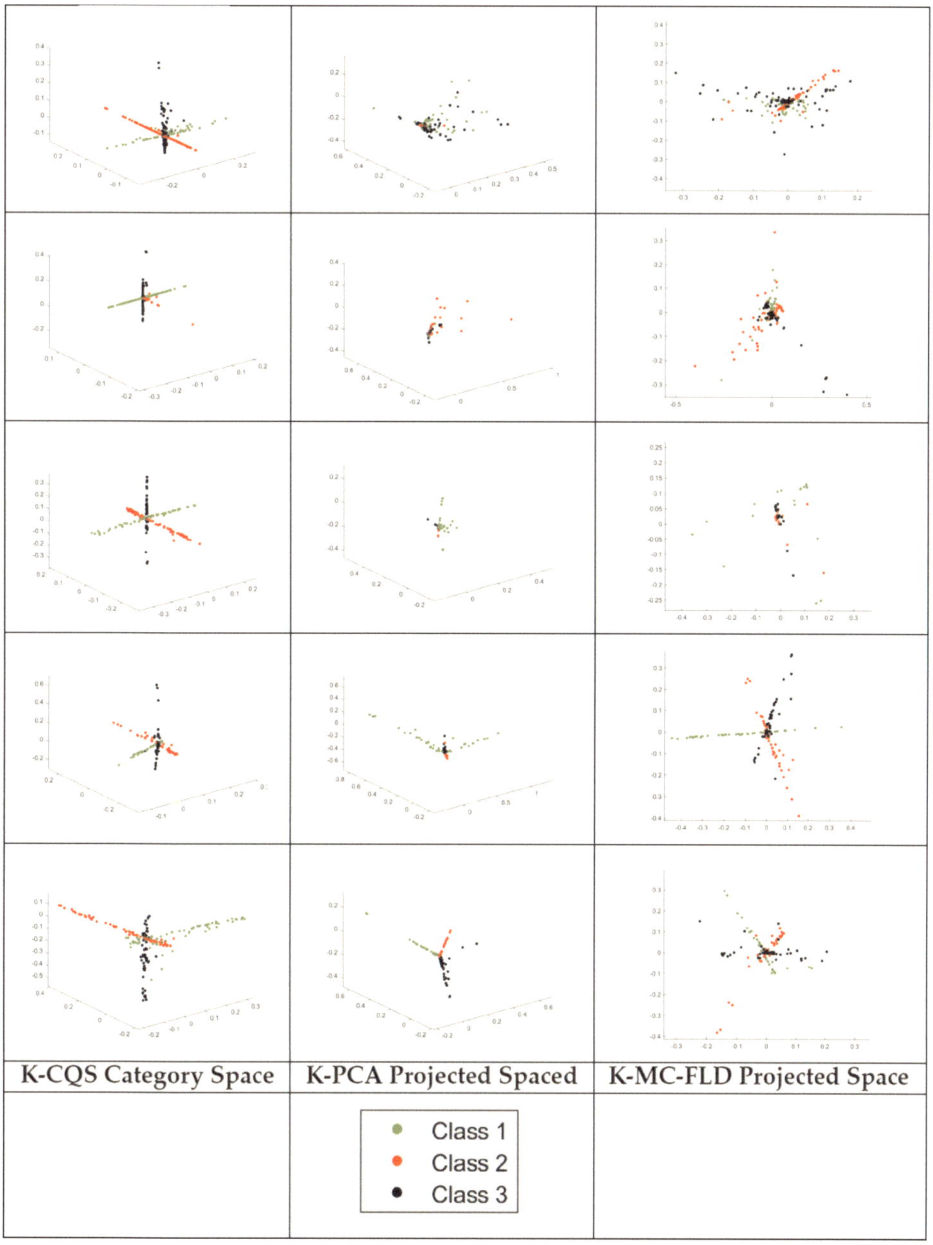

Figure 1.
Reduced dimensionality projection for a medium σ value: From top to bottom: Row 1: Vertebral, Row 2: Thyroid, Row 3: Wine, Row 4: Iris, Row 5: Seeds.

motivation for this work stems from the observation that the semantics of the multi-class Fisher linear discriminant are unclear, especially w.r.t. defining a space for the categories (classes). Beginning with this observation, we designed an objective function comprising sums of quadratic and absolute value functions (aimed at maximizing the inner product between each training set pattern and its class axes) with Stiefel manifold constraints (since the category axes are orthonormal). It turns out that recent work has characterized such problems and provided sufficient conditions for the detection of

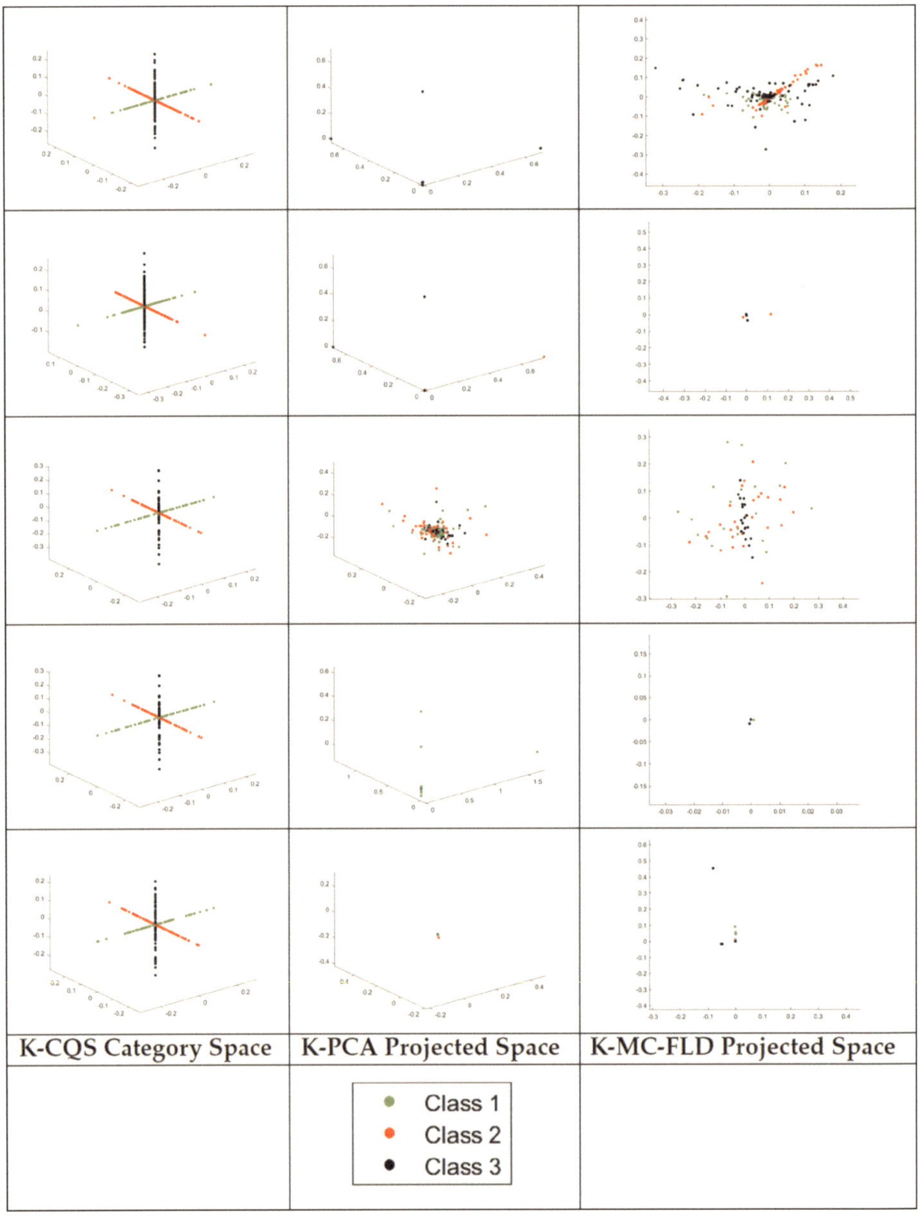

Figure 2.
Reduced dimensionality projection for a small σ value. From top to bottom: Row 1: Vertebral, Row 2: Thyroid, Row 3: Wine, Row 4: Iris, Row 5: Seeds.

global minima (despite the presence of non-convex constraints). The availability of a straightforward Stiefel manifold optimization algorithm tailored to this problem (which has no step size parameters to estimate) is an attractive by-product of this formulation. The extension to the kernel setting is entirely straightforward. Since the kernel dimensionality reduction approach warps the patterns toward orthogonal category axes, this raises the possibility of using the angle between each pattern and the category axes as a

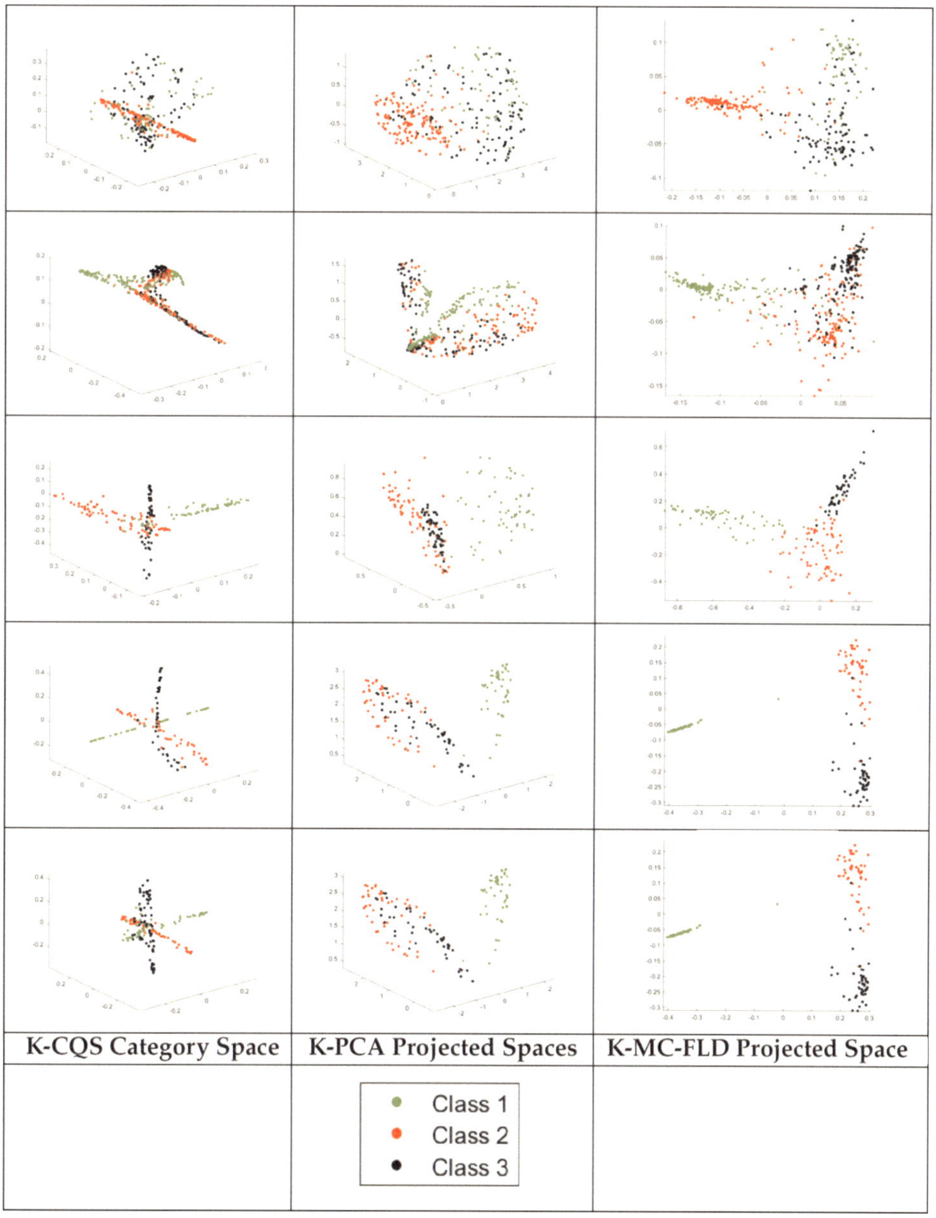

Figure 3.
Reduced dimensionality projection for a large σ value. From top to bottom: Row 1: Vertebral, Row 2: Thyroid, Row 3: Wine, Row 4: Iris, Row 5: Seeds.

classification measure. We conducted experiments in the kernel setting and demonstrated reasonable performance for the angle-based classifier, suggesting a new avenue for future research. Finally, visualization of dimensionality reduction for three classes showcases the category space geometry with clear semantic advantages over standard principal components and multi-class Fisher.

Several opportunities exist for future research. We notice a clustering of patterns near the origin of the category space, clearly calling for an origin margin (as in support vector machines) [43]. At the same time, we can also remove the orthogonality assumption (in the linear case) while continuing to pursue multi-class discrimination. Finally, extensions to the multi-label case [42] are warranted and suggest interesting opportunities for future work.

Nomenclature

Name	Description
PCA	principal component analysis
SVM	support vector machine
FLD	Fisher linear discriminant
CCA	canonical correlation analysis
PLS	partial least squares
SPCA	supervised PCA
RKHS	reproducing Kernel Hilbert space
KKT	Karush-Kuhn-Tucker
CQS	category quadratic space
CAS	category absolute value space
LS-LDA	least squares linear discriminant analysis
MC-FLD	multi-class Fisher linear discriminant
K-CQS	kernel category quadratic space
K-CAS	kernel category absolute value space
K-PCA	kernel principal component analysis
K-MC-FLD	kernel multi-class Fisher linear discriminant
K-CQS-A	kernel CQS w/angle classification
K-CAS-A	kernel CAS w/angle classification

Additional information

An earlier version of this chapter was previously published as a preprint in: 1. Smith AO, Rangarajan A. A Category Space Approach to Supervised Dimensionality Reduction [Internet]. arXiv.org. 2016 [cited 2023 Sep 5]. Available from: https://arxiv.org/abs/1610.08838

Author details

Anthony O. Smith[1]* and Anand Rangarajan[2]

1 Department of Electrical and Computer Engineering, Florida Institute of Technology, Melbourne, FL, USA

2 Department of Computer and Information Science and Engineering, University of Florida, Gainesville, FL, USA

*Address all correspondence to: anthonysmith@fit.edu

References

[1] Belhumeur PN, Hespanha JP, Kriegman DJ. Eigenfaces vs. Fisherfaces: Recognition using class specific linear projection. IEEE Transactions on Pattern Analysis and Machine Intelligence. 1997; **19**(7):711-720

[2] Rapcsák T. On minimization on Stiefel manifolds. European Journal of Operational Research. 2002;**143**(2): 365-376

[3] Jiang B, Dai Y-H. A framework of constraint preserving update schemes for optimization on Stiefel manifold. Mathematical Programming. 2015; **153**(2):535-575

[4] Bolla M, Michaletzky G, Tusnady G, Ziermann M. Extrema of sums of heterogeneous quadratic forms. Linear Algebra and its Applications. 1998;**269** (1–3):331-365

[5] Liu H, Wu W, So AM-C. Quadratic optimization with orthogonality constraints: Explicit Lojasiewicz exponent and linear convergence of line-search methods. In: Proceedings of Machine Learning Research, International Conference on Machine Learning. PMLR; 2016. pp. 1158-1167. Available from: https://proceedings.mlr.press/v48/liue16.html

[6] Hardoon DR, Szedmak SR, Shawe-Taylor JR. Canonical correlation analysis: An overview with application to learning methods. Neural Computation. 2004; **16**(12):2639-2664

[7] Xu M, Zhu Z, Zhang X, Zhao Y, Li X. Canonical correlation analysis with l 2, 1-norm for multiview data representation. IEEE Transactions on Cybernetics. 2019; **50**(11):4772-4782

[8] Roweis ST, Saul LK. Nonlinear dimensionality reduction by locally linear embedding. Science. 2000; **290**(5500):2323-2326

[9] Chen J, Liu Y. Locally linear embedding: A survey. Artificial Intelligence Review. 2011;**36**:29-48

[10] Ghojogh B, Ghodsi A, Karray F, Crowley M. Locally linear embedding and its variants: Tutorial and survey. arXiv. preprint arXiv: 2011.10925. 2020

[11] Tenenbaum JB, de Silva V, Langford JC. A global geometric framework for nonlinear dimensionality reduction. Science. 2000;**290**(5500): 2319-2323

[12] Jenkins OC, Matarić MJ. A spatio-temporal extension to isomap nonlinear dimension reduction. In: Proceedings of the Twenty-First International Conference on Machine Learning. 2004. p. 56

[13] Belkin M, Niyogi P. Laplacian eigenmaps for dimensionality reduction and data representation. Neural Computation. 2003;**15**(6):1373-1396

[14] Li B, Li Y-R, Zhang X-L. A survey on Laplacian eigenmaps based manifold learning methods. Neurocomputing. 2019;**335**:336-351

[15] Zhu H, Koniusz P. Generalized Laplacian eigenmaps. Advances in Neural Information Processing Systems. 2022;**35**:30783-30797

[16] Jolliffe IT. Principal Component Analysis. Springer Series in Statistics. 2nd ed. New York: Springer; 2002

[17] Abdi H, Williams LJ. Principal component analysis. Wiley Interdisciplinary Reviews: Computational Statistics. 2010;**2**(4):433-459

[18] Bro R, Smilde AK. Principal component analysis. Analytical Methods. 2014;**6**(9):2812-2831

[19] Fisher RA. The use of multiple measurements in taxonomic problems. Annals of Eugenics. 1936;7(2):179-188

[20] Hotelling H. Analysis of a complex of statistical variables into principal components. Journal of Educational Psychology. 1933;**24**(6):417-441

[21] Arenas-García J, Petersen KB, Hansen LK. Sparse kernel orthonormalized PLS for feature extraction in large data sets. Advances in Neural Information Processing Systems. 2007;**19**:33-40

[22] Vinzi VE, Chin WW, Henseler J, Wang H, et al. Handbook of Partial Least Squares. Vol. 201. Springer; 2010. DOI: 10.1007/978-3-540-32827-8

[23] Vapnik VN. Statistical Learning Theory. John Wiley & Sons; 1998. Available from: https://www.wiley.com/en-us/Statistical+Learning+Theory-p-978047103003

[24] Bishop CM. Neural Networks for Pattern Recognition. 1st ed. Oxford University Press; 1996. Available from: https://global.oup.com/academic/product/neural-networks-for-pattern-recognition-9780198538646?cc=us&lang=en&

[25] Duda RO, Hart P, Stork DG. Pattern Classification. 2nd ed. New York, NY: Wiley Interscience; 2000

[26] Hastie T, Tibshirani R. Discriminant analysis by Gaussian mixtures. Journal of the Royal Statistical Society, Series B (Methodological). 1996;**58**(1):155-176

[27] Ye J. Least squares linear discriminant analysis. In: Proceedings of the 24th International Conference on Machine Learning (ICML). ACM; 2007. pp. 1087-1093. DOI: 10.1145/1273496.1273633

[28] Bishop CM. Pattern Recognition and Machine Learning. 1st ed. New York: Springer; 2006

[29] Sammon JW. An optimal discriminant plane. IEEE Transactions on Computers. 1970;**100**(9):826-829

[30] Foley DH, Sammon JW. An optimal set of discriminant vectors. IEEE Transactions on Computers. 1975;**100**(3):281-289

[31] Anderson TW, Bahadur RR. Classification into two multivariate normal distributions with different covariance matrices. The Annals of Mathematical Statistics. 1962;**33**(2):420-431

[32] Schölkopf B, Burges CJC. Advances in Kernel Methods: Support Vector Learning. MIT Press; 1999. DOI: 10.7551/mitpress/1130.001.0001

[33] Rao CR. The use and interpretation of principal component analysis in applied research. Sankhyā: The Indian Journal of Statistics, Series A. 1964;**26**(4):329-358

[34] Bair E, Tibshirani R. Semi-supervised methods to predict patient survival from gene expression data. PLoS Biology. 2004;**2**(4):e108

[35] Bair E, Hastie T, Paul D, Tibshirani R. Prediction by supervised principal components. Journal of the American Statistical Association. 2012;**101**(473):119-137

[36] Widdows D. Geometry and Meaning. Vol. 773. Stanford: CSLI Publications; 2004

[37] Widdows D. Orthogonal negation in vector spaces for modelling word-meanings and document retrieval. In: Proceedings of 41st Annual Meeting on Association for Computational Linguistics. Vol. 1. Association for Computational Linguistics; 2003. pp. 136-143. DOI: 10.3115/1075096.1075114

[38] Tsochantaridis I, Hofmann T, Joachims T, Altun Y. Support vector machine learning for interdependent and structured output spaces. In: Proceedings of the Twenty-First International Conference on Machine Learning (ICML). ACM; 2004. p. 104. DOI: 10.1145/1015330.1015341

[39] Ji S, Ye J. Linear dimensionality reduction for multi-label classification. In: Proceedings of the 21st International Joint Conference on Artificial Intelligence (IJCAI). Vol. 9. 2009. pp. 1077-1082. DOI: 10.5555/1661445.1661617

[40] Johnson RA, Wichern DW. Applied Multivariate Statistical Analysis. 6th ed. Pearson; 2002. DOI: 10.1007/978-3-662-45171-7

[41] Sun L, Ji S, Ye J. Canonical correlation analysis for multilabel classification: A least-squares formulation, extensions, and analysis. IEEE Transactions on Pattern Analysis and Machine Intelligence. 2011;**33**(1):194-200

[42] Sun L, Ji S, Ye J. Multi-Label Dimensionality Reduction. CRC Press; 2013. DOI: 10.1201/b16017

[43] Shajari H, Rangarajan A. A unified framework for multiclass and multilabel support vector machines. CoRR, abs/2003.11197. 2020. DOI: 10.48550/arXiv.2003.11197

[44] Gao W, Ma Z, Xiong C, Gao T. Dimensionality reduction of spd data based on riemannian manifold tangent spaces and local affinity. Applied Intelligence. 2023;**53**(2):1887-1911

[45] Ghojogh B, Crowley M, Karray F, Ghodsi A. Elements of Dimensionality Reduction and Manifold Learning. Springer Nature; 2023. DOI: 10.1007/978-3-031-10602-6

[46] Yuille AL, Rangarajan A. The concave-convex procedure. Neural Computation. 2003;**15**(4):915-936

[47] Kiefer J. Sequential minimax search for a maximum. Proceedings of the American Mathematical Society. 1953;**4**(3):502-506

[48] Rapcsák T. On minimization of sums of heterogeneous quadratic functions on Stiefel manifolds. In: Migdalas A, Pardalos PM, Värbrand P, editors. From Local to Global Optimization. Springer; 2001. pp. 277-290. DOI: 10.1007/978-1-4757-5284-7_12

[49] Kelly M, Longjohn R, Nottingham K. The UCI Machine Learning Repository. 2013. Available from: http://archive.ics.uci.edu

Chapter 3

Application of Complex Hilbert Principal Component Analysis to the Economic Phenomena

Wataru Souma

Abstract

Conventional principal component analysis operates using a correlation matrix that is defined in the space of real numbers. This study introduces a novel method—complex Hilbert principal component analysis—which analyzes data using a correlation matrix defined in the space of complex numbers. As a practical application, we examine 10 major categories from the Japanese Family Income and Expenditure Survey for the period between January 1, 2000 and June 30, 2023, paying special attention to the time periods preceding and following the onset of the novel coronavirus disease 2019 pandemic. By analyzing the mode signal's peaks, we identify specific days that exhibit characteristics that are consistent with the events occurring before and after the pandemic.

Keywords: principal component analysis (PCA), complex numbers, Hilbert transformation, novel coronavirus disease 2019 (COVID-19), mode signal

1. Introduction

The analysis of big data may reveal novel aspects of nature and of our society. However, the significance of datasets is often obscured by noise, so distinguishing meaningful signals from such noise is an essential task. Principal component analysis (PCA) is a valuable tool for understanding the characteristics of a dataset by reducing its dimensionality to fewer dimensions than exist in the original dataset. PCA has long been exploited in various fields of study ranging from natural science to the social sciences and was recently used with machine learning as a dimensionality reduction method.

A pioneering economic study that used PCA by Connor and Korajczyk [1] to derive factors from a large set of stock returns, which led to PCA's application in various areas of economics from microeconomics to macroeconomics (e.g., [2–7]).

PCA's key point is to distinguish between significant and random components. The Scree criterion for factor retention introduced by Cattell [8] is useful for visually identifying the separation point. In statistical physics, the distribution of eigenvalues [9] and of components of corresponding eigenvectors [10] were derived analytically in research on random matrix theory (RMT). The authors from references [11–13] developed an RMT-based "null hypothesis" test that explicitly compares the properties of empirical equal-time cross-correlation matrices to those of random matrices. Any deviations from the properties of a random matrix were considered indicators of meaningful information. The RMT method has been applied to the equal-time cross-correlation matrix of assets [14–19].

Plerou et al. [13] constructed a "filtered" cross-correlation matrix using eigenvalues and eigenvectors outside of the bounds of a random matrix and applied it to Markowitz's portfolio optimization [20]. Their results showed that predicted risk is more closely aligned with realized risk when using a cross-correlation matrix than when using traditional portfolio optimization methods. The authors from references [21, 22] applied Markowitz's portfolio optimization technique to stocks listed in the first division of the Tokyo Stock Exchange and demonstrated that the performance of portfolios constructed using this method generally outperformed market indices such as the Tokyo Price Index (TOPIX).

RMT is a robust technique for isolating the meaningful components from background noise in financial time-series data. The null hypothesis used in this method assumes both cross-correlation and autocorrelation randomness. However, autocorrelation in stock returns is generally not random, as evidenced by studies such as in Ref. [23]. Therefore, a new method that maintains autocorrelation while randomizing cross-correlation is required. To address this need, the authors from references [24, 25] introduced Rotational Random Shuffling (RRS), a method in which empirical time-series data are rotationally shuffled along the time axis, subject to periodic boundary conditions. Equal-time cross-correlation matrices generated from RRS time-series data preserve most of the autocorrelation while effectively randomizing the cross-correlation. A comparison between the eigenvalue distribution of the RRS-generated cross-correlation matrix and that of the empirical matrix enables one to effectively differentiate between meaningful components and noise.

Extending the application of RMT to cross-correlation matrices that consider different time frames was a natural development. Arai et al. [26] introduced complex Hilbert principal component analysis (CHPCA), in which the cross-correlation matrix is formulated in complex space, and the components of the eigenvectors are distributed across the complex plane, allowing for the identification of lead-lag relationships between components using angular differences. The authors from references [27, 28] applied CHPCA to a time-series dataset of assets in the S&P 500. Vodenska et al. [29] employed CHPCA to explore foreign exchange and stock market data from 48 countries for the 1990–2012 period and revealed significant lead-lag relationships between these markets. Souma et al. [30] used CHPCA to analyze a time-series dataset for assets listed on the New York Stock Exchange from 2005 to 2014, illuminating lead-lag relationships between stocks, investment trusts, real estate investment trusts, and exchange-traded funds. The authors from references [31, 32] applied CHPCA to the early warning indicators for a financial crisis proposed by the Bank of Japan to investigate shifts in the lead-lag relationships between indices before and after a financial crisis.

When applying CHPCA to time-series data, explicitly extracting the lead-lag relationships between the different time series is essential. The authors from references [33–35] employed the Helmholtz-Hodge decomposition (HHD) to identify circular and gradient flows within complex networks. Iyetomi [36] used CHPCA and HHD on monthly time-series data for 57 U.S. Macroeconomic indicators and five trade/money indices and statistically confirmed significant co-movement among these time series, pinpointing significant economic events. Iyetomi et al. [37] summarized CHPCA, RRS, and HHD and applied these methodologies to economic time-series data.

Souma et al. [38] investigated the complex interdependencies between the economic policy uncertainty (EPU) and geopolitical risk (GPR) indices across 31 countries by using CHPCA to identify key events that prompted significant changes in EPU and GPR index values for the 1997–2020 period and examining the leading and lagging relationships between countries to determine whether one country's EPU or GPR index could influence those of another's. This study demonstrated that CHPCA enables a weighted and directed network to be constructed from the correlation matrix and concluded that the most impactful event of this period was the September 11, 2001 terrorist attacks, followed by the COVID-19 pandemic in 2020 and the global financial crisis in 2008.

The objective of this chapter is to apply CHPCA to the Japanese Family Income and Expenditure Survey (FIES), which is conducted by the Statistics Bureau of Japan [39], and to clarify the characteristics of consumption before and after the COVID-19 pandemic. In Section 2, we describe the characteristics of the FIES data, which consists of daily consumption records extending from January 1, 2000 through June 30, 2023. We remove trends, seasonality, and weekly effects from this dataset. In Section 3, we outline the CHPCA methodology and discuss random shuffling (RS), RRS, and mode signals. In Section 4, we present our findings, demonstrating that CHPCA is able to successfully extract the characteristics of typical events occurring before and after the COVID-19 pandemic using consumption as the mode signal. Finally, Section 5 presents a summary and discussion.

2. Data

This chapter investigates the FIES conducted by the Statistics Bureau of Japan [39]. The FIES is a sample survey that includes households, excluding single-person student households. The 2015 Population Census covered 51.57 million households and represented 96.5% of all Japanese households. Sample households for the FIES are chosen using statistical methodologies that ensure that the FIES data may represent all households in the country. Approximately 9000 households were selected using a three-stage sampling method. Each sampled family household recorded its income and expenditure in a family account ledger for 6 months and for 3 months single-person households.

The FIES comprises multi-person households (i.e., those with two or more persons) and single-person households. Of the approximately 9000 households surveyed, about 85.7% are multi-person households, while approximately 14.3% are single-person households.

Although the proportion of single-person households has recently been increasing, this chapter focuses on data for multi-person households, as they remain the predominant type of Japanese household. The FIES began in September 1950, and census data starting in January 2000 can be downloaded from reference [40]. The number of disclosed items exceeds 150. There are 10 major consumption expenditure categories as follows: 1. Food, 2. Housing, 3. Fuel, light & water charges, 4. Furniture & household utensils, 5. Clothing & footwear, 6. Medical care, 7. Transportation & communication, 8. Education, 9. Culture & recreation, and 10. Other Consumption Expenditures.

While FIES reports and statistical tables are published monthly, they feature daily data, so this section delves into the characteristics of the daily data associated with the

10 primary consumption items. The dataset includes six leap years. To standardize each year to 365 days for this analysis, we compute the average consumption expenditures for February 28 and 29 and then assign this average to February 28. Then, we exclude February 29. We denote the consumption expenditure of the n-th item in year y on day d as $X_{n,y,d}$.

Figure 1 portrays the daily fluctuations of these 10 primary consumption items from January 1, 2000, to June 30, 2023. In this figure, each white line represents the one-year moving average, highlighting the existence of a trend. To remove the trend from each time series, we calculate the change relative to the corresponding day in the previous year using the following equation:

$$R_{n,y,d} = \frac{X_{n,y,d}}{X_{n,y-1,d}}. \tag{1}$$

Hereafter, we will denote $R_{n,y,d}$ as $R_{n,t}$.The results of this transformation are illustrated in **Figure 2**. In this figure, each white line represents a one-year moving average of $R_{n,t}$, which highlights the existence of a trend denoted $G_{n,t}$.

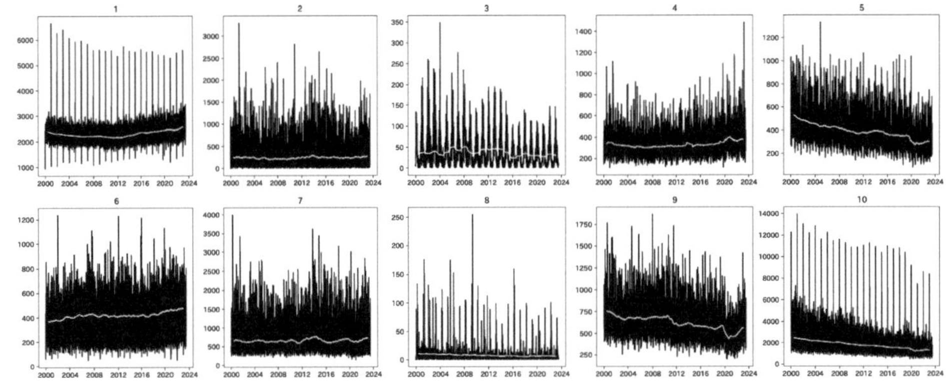

Figure 1.
Original data for 10 categories of the Japanese family income and expenditure survey. Source: The statistics Bureau of Japan.

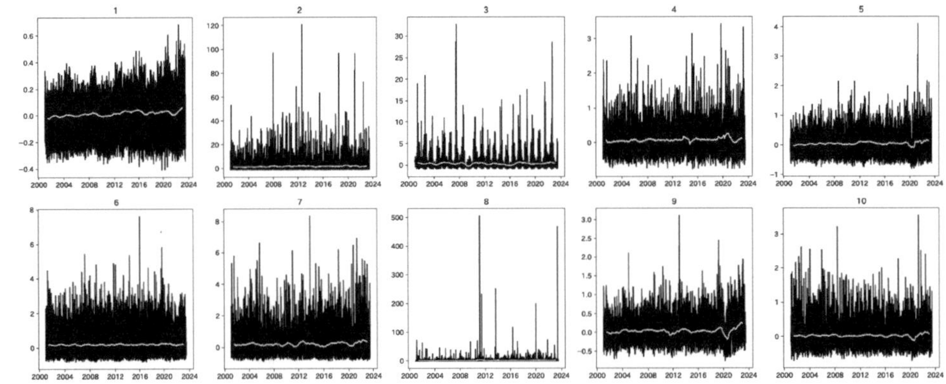

Figure 2.
Change in 10 categories of the Japanese family income and expenditure survey from the same day the previous year.

In addition to the trend $G_{n,t}$, the time series $R_{n,t}$ contains seasonality $S_{n,t}$. To remove these components, we obtain the following equation:

$$Y_{n,t} = R_{n,t} - G_{n,t} - S_{n,t}. \tag{2}$$

However, $Y_{n,t}$ is still subject to day-of-the-week effects, $D_{n,t}$, defined as follows:

$$D_{n,t} = \frac{1}{N_d} \sum_{t \in d} Y_{n,t}, \tag{3}$$

where N_d represents the number of days in a week. Therefore, after removing $D_{n,t}$ from $Y_{n,t}$, we obtain the following:

$$r_{n,t} = Y_{n,t} - D_{n,d}. \tag{4}$$

The resulting time series passes both the Augmented Dickey-Fuller test and the Kwiatkowski-Phillips-Schmidt-Shin tests.

3. Methods and materials

In this section, we explain how CHPCA is used to explore the correlation structure of $r_{n,t}$.

3.1 Complex correlation matrix

The mean value of $r_{n,t}$ is defined as follows:

$$\langle r_n \rangle = \frac{1}{T} \sum_{t=1}^{T} r_{n,t}. \tag{5}$$

The variance is given by the following:

$$\sigma_n^2 = \frac{1}{T} \sum_{t=1}^{T} \left(r_{n,t} - \langle r_n \rangle \right)^2. \tag{6}$$

To standardize $r_{n,t}$, as defined in Eq. (4), to have a mean of zero and a variance of one, we use the following:

$$w_{n,t} = \frac{r_{n,t} - \langle r_n \rangle}{\sigma_n}. \tag{7}$$

The Fourier transform of Eq. (7) is given as follows:

$$w_{n,t} = \sum_{k=0}^{T} [a_n(\omega_k) \cos(\omega_k t) + b_n(\omega_k) \sin(\omega_k t)], \tag{8}$$

where $\omega_k = 2\pi k / T \geq 0$.

The Hilbert transform of Eq. (8) is given by the following:

$$\hat{w}_{n,t} = \sum_{k=0}^{T} [b_n(\omega_k)\cos(\omega_k t) - a_n(\omega_k)\sin(\omega_k t)]. \tag{9}$$

Eq. (9) corresponds to Eq. (8) shifted to the phase $\pi/2$. Therefore, Eqs. (8) and (9) are orthogonal to each other. Now, using Eqs. (8) and (9), we define the complex time series as follows:

$$\tilde{w}_{n,t} = w_{n,t} + i\hat{w}_{n,t} = \sum_{k=0}^{T} c_n(\omega_k)e^{-i\omega_k t}, \tag{10}$$

where i is an imaginary unit defined by $i^2 = -1$, and $c_n(\omega_k) = a_n(\omega_k) + ib_n(\omega_k)$.

The right-hand side of Eq. (10) indicates that $\tilde{w}_{n,t}$ rotates in a clockwise direction over time. The matrix with the components specified by Eq. (10) is known as the complex Wishart matrix, and it can be expressed as follows:

$$\tilde{\mathbf{W}} = [\tilde{w}_{n,t}]. \tag{11}$$

Consequently, the complex correlation matrix is defined by the following:

$$\mathrm{C} = \frac{1}{T}\tilde{\mathbf{W}}\tilde{\mathbf{W}}^{\dagger}, \tag{12}$$

where $\tilde{\mathbf{W}}^{\dagger}$ denotes the adjoint matrix of $\tilde{\mathbf{W}}$ (i.e., the transpose and complex conjugate).

The components of the complex correlation matrix can be expressed as follows:

$$C_{mn} = \mathrm{Re}(C_{mn}) + i\mathrm{Im}(C_{mn}), \tag{13}$$

$$= |C_{mn}|e^{i\varphi_{mn}}, \tag{14}$$

where φ_{mn} represents the correlation in phase space. The leading or lagging behavior of each component is determined using φ_{mn}.

3.2 Random shuffling and rotational random shuffling

We can generate a random complex correlation matrix devoid of both autocorrelation and cross-correlation by employing a randomly shuffled Wishart matrix as follows:

$$w_{n,t} \rightarrow w_{n,\mathrm{rand}[1,T]}, \tag{15}$$

where rand$[1, T]$ denotes a random integer between 1 to T selected without repetition. This procedure is referred to as RS.

Many economic time series display autocorrelation, so a method that retains autocorrelation while randomizing cross-correlation is helpful. To achieve this, we create a random complex correlation matrix that preserves autocorrelation but eliminates cross-correlation, which can be accomplished by using a rotationally and randomly shuffled Wishart matrix, defined as follows:

$$w_{n,t} \to w_{n,\mathrm{mod}[t+\tau,T]}, \tag{16}$$

where $\tau \in [0, T-1]$ is a pseudo-random integer unique to each n. This procedure is referred to as RRS. It should be noted, however, that the imposition of a periodic boundary condition disrupts autocorrelation at that point.

3.3 Decomposition of the complex correlation matrix

When PCA is applied to the complex correlation matrix, **C**, it yields the eigenvalue λ_j and its corresponding eigenvector $\mathbf{v}_j$, where j denotes the ranking of the eigenvalues and their corresponding eigenvectors. By employing PCA, if we determine the number of principal components as n_p, we may decompose the complex correlation matrix into its meaningful and noisy parts as follows:

$$\mathbf{C} = \sum_{j=1}^{N} \lambda_j \boldsymbol{v}_j \boldsymbol{v}_j^\dagger = \sum_{j=1}^{n_\mathrm{p}} \lambda_j \boldsymbol{v}_j \boldsymbol{v}_j^\dagger + \sum_{j=n_\mathrm{p}+1}^{N} \lambda_j \boldsymbol{v}_j \boldsymbol{v}_j^\dagger = \mathbf{P} + \mathbf{R}, \tag{17}$$

where $\boldsymbol{v}_j^\dagger$ represents the adjoint vector of $\boldsymbol{v}_j$, which is its transformation and complex conjugate. In Eq. (17), **P** and **R** denote the principal and noisy components of the complex correlation matrix, respectively. Thus, examining **P** is a logical means for uncovering the characteristics of the correlation.

3.4 Mode signal

A mode signal, denoted α_j, is a vector the number of components of which is equal to the length of the time series, T. It is defined by the product of $\boldsymbol{v}_j$ and **W** as follows:

$$\alpha_j = \boldsymbol{v}_j^\dagger \mathbf{W}. \tag{18}$$

The mode signal serves as a useful tool for identifying the sympathetic structure of the time series.

4. Results

In Section 2, we discussed 10 major categories of consumption expenditure. However, the categories of Housing, Fuel, Light & Water Charges, and Education primarily consist of monthly payment items and therefore do not accurately represent daily consumption patterns. Therefore, we excluded these three categories and focused on the remaining seven major consumption expenditure categories. We applied the methodology outlined in Section 3 to the FIES data, specifically targeting the following seven categories: 1. Food, 4. Furniture & household utensils, 5. Clothing & footwear, 6. Medical care, 7. Transportation & communication, 9. Culture & recreation, and 10. Other Consumption Expenditures.

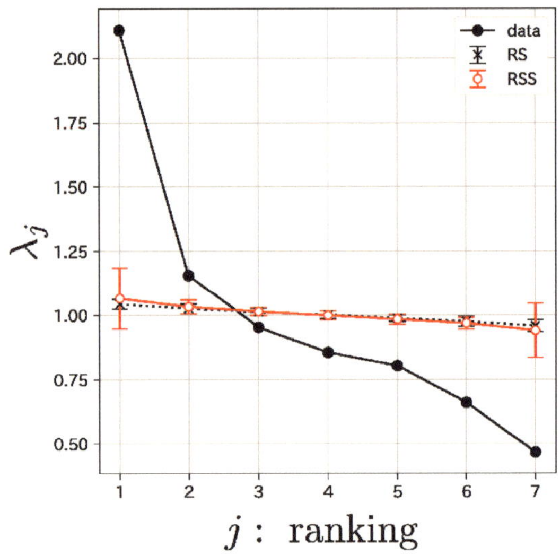

Figure 3.
Distribution of eigenvalues.

4.1 Eigenvalues

Figure 3 presents the scree graph for the eigenvalues where the abscissa represents the ranking, j, of each eigenvalue, while the ordinate indicates the magnitude of each eigenvalue λ_j.

The black line marked with filled circles corresponds to the distribution of eigenvalues obtained from the complex correlation matrix constructed from the data. The black dotted line marked with crosses and accompanied by error bars represents the distribution of the eigenvalues sourced from the complex correlation matrix generated using RS. The red line marked with open circles and error bars depicts the distribution of eigenvalues derived from the complex correlation matrix generated using RRS. We conducted 50 simulations for RS and RRS to determine their respective mean values and standard deviations. In this graph, the error bars represent three times the standard deviation.

By comparing these results, we confirm that the first eigenvalue is evidently the principal component, that is, $n_{\mathrm{p}} = 1$.

4.2 Principal eigenvector

Figure 4 illustrates the distribution of the eigenvector that corresponds to the first eigenvalue. In this figure, the abscissa represents the real axis, while the ordinate represents the imaginary axis. Unlike the components in a real correlation matrix, the components of this eigenvector are distributed on a complex plane. To account for the nature of the eigenvectors in this figure, we imposed a constraint that sets the imaginary part of "10. Other Consumption Expenditures" to zero.

In this context, two key quantities are the absolute value and the argument of each complex component. The absolute value signifies the strength of each component in the eigenvector. Therefore, "10. Other Consumption Expenditures" emerges as the most dominant factor, followed by "1. Food."

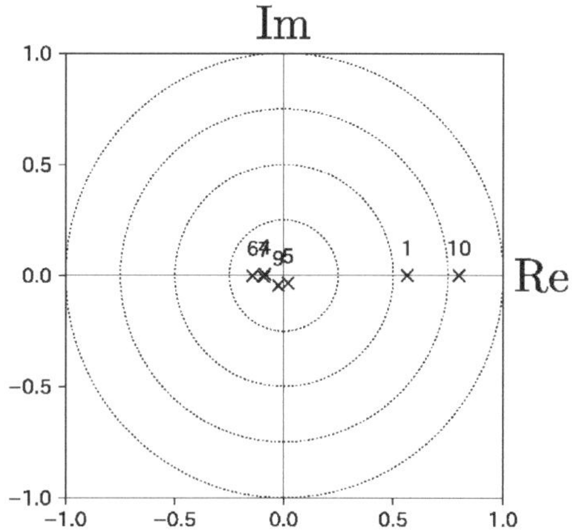

Figure 4.
Distribution of components of principal eigenvector.

The argument of the complex number indicates the leading and lagging relationships between the factors. As demonstrated in Eq. (10), each index rotates in a clockwise manner as time progresses. Nevertheless, **Figure 4** reveals no typical leading or lagging relationships. Instead, the figure identifies three distinct groups: one consisting of "10. Other Consumption Expenditures" and "1. Food"; another comprising "4. Furniture & Household Utensils," "6. Medical Care," and "7. Transportation & Communication"; and a final group containing "5. Clothing & Footwear" and "9. Culture & Recreation."

4.3 Mode signals

We consider the squared values of the mode signals defined by using the following expression:

$$|\alpha_t|^2 = \sum_{j \in J} |\alpha_{j,t}|^2, \tag{19}$$

where $J = \{1,4,5,6,7,9,10\}$ represents the numbers corresponding to the various categories from the included in our analysis. We square the mode signals because their components consist of complex numbers. **Figure 5** displays the mode signal as it is defined in Eq. (19).

In this figure, the abscissa represents the days, and the ordinate represents the squared values of the mode signals. Here, we focus on the peaks occurring after 2019. The top five peaks are summarized in **Table 1**, ordered according to time. The most significant peak occurred on September 30, 2019. This is the day before Japan's consumption tax was increased from 8% to 10%, October 1, 2019. The second peak falls on October 12, 2020, coinciding with the Japanese government's "Go To Campaign" to stimulate the economy after the COVID-19 pandemic, which had decreased consumption. The third peak is observed on April 24, 2021, when Japan was in the midst of the fourth wave of its COVID-19 pandemic; a state of emergency was declared in four prefectures in the Tokyo metropolitan area on April 25, 2021. The fourth peak occurs on May 6, 2021 and

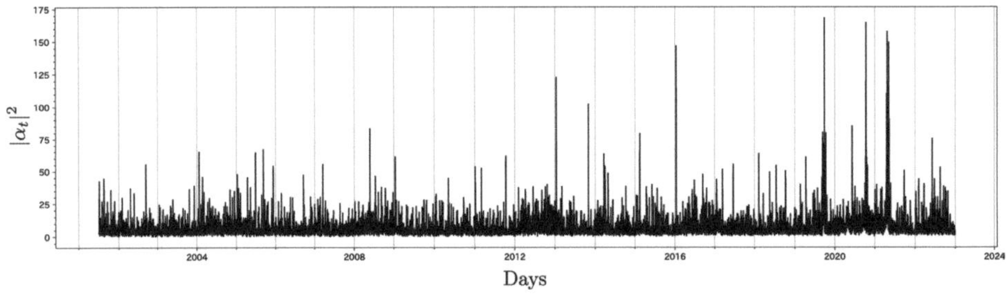

Figure 5.
The sum of mode signals $|\alpha_t|^2$ for the Japanese family income and expenditure survey data.

Data	Ranking of mode signal	Relating events
2019-09-30	1	Sales tax rate increases from 8–10% (October 1, 2019)
2020-10-12	2	Economic stimulus package for the COVID-19 pandemic
2021-04-18	5	State of emergency declared in four prefectures
2021-04-24	3	State of emergency declared in four prefectures (April 25, 2021)
2021-05-06	4	State of emergency declared in nine prefectures (May 12, 2021)

Table 1.
Five peaks of the sum of mode signals $|\alpha_t|^2$ after 2019 and the occurrence of related events.

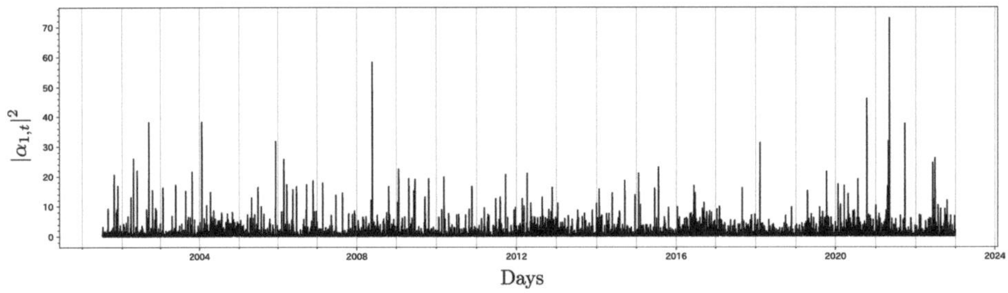

Figure 6.
The first mode signal $|\alpha_{1,t}|^2$ for the Japanese family income and expenditure survey data.

Data	Ranking of mode signal	Relating events
2019-09-30	5	Sales tax rate increases from 8–10% (October 1, 2019)
2020-10-12	2	Economic stimulus package for the COVID-19 pandemic
2021-04-26	3	State of emergency declared in four prefectures (April 25, 2021)
2021-05-06	1	State of emergency declared in nine prefectures (May 12, 2021)
2021-09-24	4	All emergency declarations lifted (September 30, 2021)

Table 2.
Five peaks of the first mode signal $|\alpha_{1,t}|^2$ after 2019 and relating events.

corresponds to the declaration of a state of emergency for nine prefectures on May 12, 2021. Finally, the fifth peak occurs on April 18, 2021 and is attributed to the same factors as the third peak.

Next, we focus on the first mode signal $|\alpha_{1,t}|^2$. **Figure 6** displays the mode signal of $|\alpha_{1,t}|^2$. In this figure, the abscissa represents the days, and the ordinate represents $|\alpha_{1,t}|^2$. Here, we once again focus on the peaks occurring after 2019. The top five peaks are summarized in **Table 2**, Unlike the data in **Table 1**, the most prominent peak for this mode signal is May 6, 2021 when Japan was in the midst of the fourth wave of its COVID-19 pandemic: A state of emergency had been declared for four prefectures in the Tokyo metropolitan area on April 25, 2021 and was extended to nine prefectures on May 12, 2021. The second peak falls on October 12, 2020, coinciding with the Japanese government's "Go To Campaign" aimed at stimulating the economy due to the decline in consumption precipitated by the pandemic. The third peak is observed on April 26, 2021, 1 day after a state of emergency was declared for four prefectures in the Tokyo metropolitan area on April 25, 2021. The fourth peak occurs on September 24, 2021, a date not exhibited in **Table 1**, corresponding to the lifting of all emergency declarations on September 30, 2021. It should be noted that people had more opportunities to go out around this time, leading to increased consumption. Finally, the fifth peak occurs on September 30, 2019, the day before Japan's consumption tax was increased from 8–10%, October 1, 2019, as explained in **Table 1**.

5. Conclusion

This chapter reviewed the methodology of CHPCA and examined the top 10 categories from the Japanese FIES for the January 1, 2000–June 30, 2023 period as a practical example, specifically delving into the periods preceding and following the COVID-19 pandemic. We identified characteristic days based on mode signal peaks and explored their relationship to the events occurring before and after the pandemic.

The most prominent peak in the sum of mode signals occurred on September 30, 2019, and the day before Japan's consumption tax was increased from 8–10%, October 1, 2019. Other peaks were associated with the spread of COVID-19 and government stimulus measures. Interestingly, the mode signal for the principal component, which accounted predominantly for the spread of COVID-19 and government interventions during the pandemic, ranked fifth, following the peak associated with the tax increase.

The Japanese government initiated a special fixed benefit of 100,000 yen for all citizens in June 2020 as part of the economic stimulus measures implemented during the COVID-19 pandemic. This benefit was claimed by 99% of the population by September 2020. However, this paper's CHPCA analysis did not detect that this measure had any significant impact. One explanation for this fact could be that the fixed benefit was spent gradually over several months and therefore did not cause an immediate spike in consumption. Another possibility is that the effect of the fixed benefit was too subtle to be captured by the FIES household survey.

Acknowledgements

I would like to thank Prof. H. Aoyama, Prof. Y. Fujiwara, Prof. H. Iyetomi, and Prof. H. Yoshikawa for useful discussions. This research was supported by a grant-in-aid for scientific research (KAKENHI), JSPS Grant Number 22 K04590.

Author details

Wataru Souma
Faculty of Data Science, Rissho University, Kumagaya, Japan

*Address all correspondence to: wataru.soma@gmail.com

References

[1] Connor G, Korajczyk RA. Performance measurement with the arbitrage pricing theory: A new framework for analysis. Journal of Financial Economics. 1986;**15**(3):373-394

[2] Ilmanen A. Time-varying expected returns in international bond markets. Journal of Finance. 1995;**50**(2):481-506

[3] Stock JH, Watson MW. Forecasting using principal components from a large number of predictors. Journal of the American Statistical Association. 2002; **97**(460):1167-1179

[4] Stock JH, Watson MW. Macroeconomic forecasting using diffusion indexes. Journal of Business & Economic Statistics. 2002;**20**(2):147-162

[5] Bai J, Ng S. Determining the number of factors in approximate factor models. Econometrica. 2002;**70**(1): 191-221

[6] Kose MA, Otrok C, Whiteman CH. International business cycles: World, region, and country-specific factors. The American Economic Review. 2003; **93**(4):1216-1239

[7] Bernanke BS, Boivin J, Eliasz P. Measuring the effects of monetary policy: A factor-augmented vector autoregressive (FAVAR) approach. The Quarterly Journal of Economics. 2005; **120**(1):387-422

[8] Cattell RB. The screen test for the number of factors. Multivariate Behavioral Research. 1966;**1**(2):245-276

[9] Marčenko VA, Pastur LA. Distribution of eigenvalues for some sets of random matrices. Mathematics of the USSR-Sbornik. 1967;**1**(4):457-483

[10] Porter CE, Thomas RG. Fluctuations of nuclear reaction widths. Physical Review. 1956;**104**(2):483-491

[11] Laloux L, Cizeau P, Bouchaud JP, Potters M. Noise dressing of financial correlation matrices. Physical Review Letters. 1999;**83**(7):1467-1470

[12] Plerou V, Gopikrishnan P, Rosenow B, Amaral LAN, Stanley HE. Universal and nonuniversal properties of cross correlations in financial time series. Physical Review Letters. 1999;**83**(7): 1471-1474

[13] Plerou V, Gopikrishnan P, Rosenow B, Amaral LAN, Guhr T, Stanley HE. Random matrix approach to cross correlations in financial data. Physical Review E. 2002;**65**(6):066126

[14] Utsugi A, Ino K, Oshikawa M. Random matrix theory analysis of cross correlations in financial markets. Physical Review E. 2004;**70**(2):026110

[15] Kim DH, Jeong H. Systematic analysis of group identification in stock markets. Physical Review E. 2005;**72**(4): 046133

[16] Pan RK, Sinha S. Collective behavior of stock price movements in an emerging market. Physical Review E. 2007;**76**(4): 046116

[17] Namaki A, Shirazi AH, Raei R, Jafari GR. Network analysis of a financial market based on genuine correlation and threshold method. Physica A: Statistical Mechanics and its Applications. 2011; **390**(21–22):3835-3841

[18] Namaki A, Jafari GR, Raei R. Comparing the structure of an emerging market with a mature one under global perturbation. Physica A: Statistical

Mechanics and its Applications. 2011; **390**(17):3020-3025

[19] Jamali T, Jafari GR. Spectra of empirical autocorrelation matrices: A random-matrix-theory–inspired perspective. EPL. 2015;**111**(1):10001

[20] Markowitz H. Portfolio selection. The Journal of Finance. 1952;7(1):77-91

[21] Fujiwara Y, Souma W, Murasato H, Yoon H. Application of PCA and random matrix theory to passive fund management. In: Takayasu H, editor. Practical Fruits of Econophysics. Tokyo: Springer; 2006. pp. 226-230

[22] Souma W. Toward a practical application of Econophysics: An approach from random matrix theory (written in Japanese). Applied Mathematics. 2005;**15**(3):45-59

[23] Lo AW, MacKinlay AC. An econometric analysis of nonsynchronous trading. Journal of Econometrics. 1990; **45**(1–2):181-211

[24] Iyetomi H, Nakayama Y, Aoyama H, Fujiwara Y, Ikeda Y, Souma W. Fluctuation-dissipation theory of input-output interindustrial relations. Physical Review E. 2011;**83**(1):016103

[25] Iyetomi H, Nakayama Y, Yoshikawa H, Aoyama H, Fujiwara Y, Ikeda Y, et al. What causes business cycles? Analysis of the Japanese industrial production data. Journal of the Japanese and International Economies. 2011;**25**(3):246-272

[26] Arai Y, Yoshikawa T, Iyetomi H. Complex principal component analysis of dynamic correlations in financial markets. Frontiers in Artificial Intelligence and Applications. 2013;**255**: 111-119

[27] Arai Y, Yoshikawa T, Iyetomi H. Dynamic stock correlation network. Procedia Computer Science. 2015;**60**: 1826-1835

[28] Souma W. Characteristics of principal components in stock price correlation. Frontiers in Physics. 2021;**9**: 602944

[29] Vodenska I, Aoyama H, Fujiwara Y, Iyetomi H, Arai Y. Interdependencies and causalities in coupled financial networks. PLoS One. 2016;**11**(3): e0150994

[30] Souma W, Aoyama H, Iyetomi H, Fujiwara Y, Irena V. Construction and application of new analytical methods for stock correlations: Toward the construction of prediction model of the financial crisis (written in Japanese). In: Proceeding of Network Emergent Intelligence Workshop. Tokyo: Japan Society for Software Science and Technology; 2016. pp. 1-8. Available from: http://www.ics.lab.uec.ac.jp/sig-ein/index.php?JWEIN2016

[31] Souma W, Iyetomi H, Yoshikawa H. Application of complex Hilbert principal component analysis to financial data. In: IEEE 41st Annual Computer Software and Applications Conference (COMPSAC). Vol. 2. IEEE; 2017. pp. 391-394

[32] Souma W, Iyetomi H, Yoshikawa H. The Leading and Lagging Structure of Early Warning Indicators for Detecting Financial Crises (Written in Japanese). RIETI Policy Discussion Paper Series; 18-P-005. Tokyo; 2018. pp. 1-26. Available from: https://www.rieti.go.jp/jp/publications/summary/18030017.html

[33] Kichikawa Y, Iyetomi H, Iino T, Inoue H. Hierarchical and circulating flow structure in an interfirm

transaction network. In: Abstracts of the 6th International Workshop on Complex Networks and their Applications; Lyon, France. 2017. pp. 12-14

[34] Iyetomi H, Ikeda Y, Mizuno T, Ohnishi T, Watanabe T. International trade relationship from a multilateral. In: Abstracts of the 6th International Workshop on Complex Networks and their Applications; Lyon, France. 2017. pp. 253-255

[35] Kichikawa Y, Iyetomi H, Iino T, Inoue H. Community structure based on circular flow in a large-scale transaction network. Applied Network Science. 2019;**4**(1):92. DOI: 10.1007/s41109-019-0202-8

[36] Iyetomi H. Collective phenomena in economic system. In: Complexity, Heterogeneity, and the Methods of Statistical Physics in Economics. Singapore: Springer; 2020. pp. 177-201

[37] Iyetomi H, Aoyama H, Fujiwara Y, Souma W, Vodenska I, Yoshikawa H. Relationship between macroeconomic indicators and economic cycles in US. Scientific Reports. 2020;**10**(1):1-12

[38] Souma W, Roma CM, Goto H, Iyetomi H, Vodenska I. Complex Global Interdependencies between Economic Policy Uncertainty and Geopolitical Risks Indices. RIETI Discussion Paper Series; 22-E-028. Tokyo; 2022. pp. 1-36. Available from: https://www.rieti.go.jp/en/publications/summary/22030031.html

[39] Statistics Bureau of Japan: Available from: https://www.stat.go.jp/english/data/kakei/index.html.

[40] Statistical tables (in Japanese): Available from: https://www.e-stat.go.jp/stat-search/files?page=1&layout=datalist&toukei=00200561&tstat=000000330001&cycle=1&tclass1=000000330001&tclass2=000000330004&tclass3=000000330005&tclass4val=0.

Chapter 4

Confirmatory Factor Analysis of the Life Skills Measurement Tool for University Students

Atallah Ahmed, Touati Hayat,
Saad Mohammed Abdelmoudjib, Amrani Amel,
Berrabah Ameur, Cherifi Selma, Allali Taleb and
Benkhaled Hadj

Abstract

To measure life skills, tools have been developed for this purpose, and these tools were forms to measure these variables under study and research, and through this study we would like to confirm the life skills scale through the use of confirmatory factor analysis, which is used in testing assumptions that necessarily assume the existence of patterns or special factors of relationships in the data on the basis of which variables can be classified. The individual builds a model that supposedly describes and interprets empirical data in light of relatively few parameters, which gives us a model that is more likely to measure the variables under study.

Keywords: confirmatory, factor analysis, life skills, measurement tool, university students

1. Introduction

The topic of measuring life skills is one of the topics that researchers have addressed a lot, relying on its definitions and the foundations of its construction. Life skills are considered an end in themselves that must be available in people with whom you communicate and constantly continue life, and they actively contribute to the individual's acquisition of a set of basic skills, which enable him to interact and deal with the environmental difficulties surrounding him and ensure his ability to think and make the right decision positively throughout his life, and does not stop at a certain period or at a certain age, but is constantly evolving during the stages of life and its changes, whether environmental or social. It is capable of giving the individual the privacy of positive adaptation to the situations and problems that he faces during his daily life and with which he deals positively with him and other people and situations. Thus, he is able to possess self-learning skills that enable him to learn at all times and longevity inside and outside the study place. Life skills are of particular importance, as they help in shaping and refining the personality of an individual, preparing him to face the issues of the Times and the problems of everyday life, to be a creative, productive and active person locally and globally, capable of development, development, change and local and international competition.

Hayek stated about Teo [1] in this context" that the possession of life skills provides an individual with the weapon of coexistence, adaptation, success, the ability to achieve effective communication with others and transfer what he has learned beyond the classroom [2].

Many researchers have been interested in studying and measuring the degree of acquisition of life skills among university students, i.e., higher education outputs, including the study of Sobhi Lulu and Qishta [3], Obeidat and Saada [4], Hartmann [5]. Longitudinal study on the quality of life and adjustment strategies of breast cancer patients and their "companion-referent." Magdy [6], the effectiveness of using information technologies in achieving the dimensions of quality of life among samples of Omani students, Proceedings of the Psychology and Wellbeing Symposium; Gatab et al. [7], Students' life quality prediction based on life skills, and studies aimed at finding out the positive impact of some programs on the development of life skills, such as the study of Omar [8], Al-Hayek [9]; Ayyad and Al-Din [10], Slav [11], Al-Ajmi [12]; Quality of life and its relationship to the future orientation among students of the Faculty of Graduate Studies at Naif Arab University for Security Sciences "Factor Study", El-Sherbiny and Walid [13], Hala [14]; Life Skills for People with Mental Inertia from Students of the Faculty of Arts at Al-Qadisiyah University Uruk for Humanities Somaya and Dalia [15], Muhammad [16], Mohammadi [17], Amna [2]; these studies dealt with aspects and various fields have emphasized the importance of life skills in education for the student coming to graduation, in which the educational attitude based on life skills programs is the basis in the formation, and in this regard, the Ministry of education emphasizes the importance of learning based on life skills in" as it seeks to develop the abilities of students and develop them to adapt to real life situations, and develop their thinking skills before any work or task performance to ensure a useful life, and achieve sound and positive results" [18], as Kothar kujak stressed "the need to pay attention to life skills, and provide each learner with them, so that he can face the modern changes and challenges that characterize this era, and at the same time be able to perform the work required of him to the fullest, these skills bring him successful coexistence, adaptation, flexibility and success in his work and personal life, and these skills are multiple and varied, covering all areas of life" [19, 20].

Several studies have focused on measuring life skills and have developed tools for this purpose, and these tools were forms to measure these variables under study and research, and through this study we would like to confirm the measure of life skills, and this through a psychometric study of the life skills measurement tool. Emphasis has been placed on university training. On this basis, we put forward the following: What is the factor structure of the components of the life skills measurement tool for university students using confirmatory factor analysis.

1.1 Objectives

Identify the factor structure of the components of the life skills measurement tool for university students using confirmatory factor analysis.

1.2 Hypotheses

The factor structure of the components of the life skills measurement tool for university students using confirmatory factor analysis corresponds to the proposed tool structure.

2. Terminology

2.1 Life skills

In the language you know the skill, the dexterity of the thing, and you have (mastered) the thing - stared at it [21]. And Amhara is opening too ([22], p. 561). It is also defined by a term as "the ability to perform mental, emotional, motor activity or both, and its learning or acquisition requires ease, accuracy and time economy in its performance" ([23], p. 15) and ([24], p. 240), and is also defined in the dictionary of sociology as "a complex organization of behavior developed through the learning process and the direction towards a specific goal or focus on a specific activity" ([25], p. 116). Pedagogues defined it as "a series of movements that can be observed both directly and indirectly, performed by a certain person or a number of people in the course of pursuing a goal or performing a task" ([26], p. 25). Namely, the ease and accuracy in conducting work and grow as a result of the education process.

Namely, to carry out a certain process with a degree of speed and mastery with economy in the effort expended [27]. Ahmed Zaki Saleh defined it as "ease and accuracy in performing a work with a degree of speed and mastery with economy of effort and with the least possible time by understanding" ([28], p. 123). As defined by the World Health Organization (WHO),"the ability to adopt an adaptive and positive behavior that enables to deal effectively with the demands and challenges of everyday life" ([29], p. 03). It was reported It is defined by UNICEF as psychological, social, interpersonal skills, exchange skills, scientific and professional skills that an individual needs in facilitating communication with others, negotiating with them appropriately, critical thinking and problem solving skills" [30]. It is also expressed as a set of behaviors acquired by the student that help him adapt to different and changing life requirements and face everyday problems, and is expressed in this study in the results shown by the study tool from all the above, we can conclude that skill is the ability to carry out an activity or a set of life activities, including emotional, intellectual, motor or physical, easily, masterfully and accurately in the least time, and this is to reach the desired end.

2.2 University students

They are all students who have studied for 5 years at the University and are about to obtain a master's degree in branch specialties at various universities in the National country.

2.3 Factorial analysis

Factorial analysis is a statistical technique aimed at interpreting the coefficients of positive correlations that have statistical significance between various variables. Or it is a mathematical process aimed at simplifying the correlations between the various variables involved in the analysis down to the common factors that describe the relationship between these variables and their interpretation. Therefore, factor analysis is a statistical method for analyzing multiple data that are related to each other with different degrees of correlation in the form of independent classifications based on qualitative bases of classification [31]. Factor analysis begins by calculating the correlation coefficients between a number of variables, and then we will get a matrix of correlations between these variables in the research sample on which the measurement was performed, and then this correlation matrix is followed by a factor analysis to reach the minimum possible number of axes or factors that enable us to express the greatest amount of variation between these variables.

2.4 Exploratory factor analysis

This type is used in cases where the relationships between the variables and the underlying factors are unknown and therefore the factorial analysis is aimed at discovering the factors to which the variables are described.

2.5 Confirmatory factor analysis (CFA)

It is used in testing hypotheses that necessarily assume the presence of special patterns or factors of relationships in the data on the basis of which variables can be classified. The individual builds a model that supposedly describes and interprets empirical data in light of relatively few parameters.

3. Actions

3.1 Method

The descriptive method was used in a survey style to suit the nature of the study.

3.2 Community

The research community is university students in various disciplines. We relied on (20) universities from different Algerian national countries, chosen randomly.

3.2.1 Study sample

We selected a random sample of 471 second-year master's students. Their results were analyzed in confirmatory factorial analysis.

3.2.2 Life skills tool

After reviewing a number of studies that dealt with the topic of life skills, especially the study of Imran Taghreed and others life skills, 2001, as well as Hamad Hassan and Dua [32], Fahim [33] and UNICEF [34], WHO classification 1993, adding the classification of the Center for the development of curricula and educational materials of the Ministry of education in Egypt 2000, the Ministry of education of the kingdom of Saudi Arabia Ghanem Ghanem, Saad and others. The Palestinian Ministry of education and higher education in 2003, Fouad Ayad Ismail and Hedi Bassam Saad al-Din, Amor Omar in 2008, Nayef Mufti Nahar al-Jabour in 2012, Nidal Ahmed Ismail al-Ghafri in 2012, and Tutti Hayat study in 2014, as well as another study in 2018 [27].

These studies categorized life skills into dimensions, and each dimension contains a set of questions that measure each specific life skill. We have collected a set of

Always	Often	Sometimes	Rarely	Never
05 points	04 points	03 points	02 points	01 points

Table 1.
Point of response to the question.

questions in this context that represent life skills among university students and have developed an assessment scale for them as follows:

A **Table 1** representing the answer keys to the questions of the tool.

3.2.2.1 Study limitations

Spatial limitations: The study and its instrument were applied to a sample of university students on a national level, encompassing 20 universities from the east, west, north, and south.

Human limitations: The questions were administered to a sample of 1080 students on a national level, randomly selected from master's level students.

Temporal limitations: The research was conducted during the academic year 2019/2020.

3.2.2.2 Presentation of results and discussion

A **Table 2** showing the labels for the extracted factors and the number of expressions within each.

3.2.2.3 Study results the tool by researchers

At the beginning of the exploratory factor analysis for the life skills assessment tool, and after completing the initial process for the core components of the factors that make up the tool, we wanted to know if these results from the exploratory factor analysis would apply to the confirmatory factor analysis using a different sample than the one used initially.

After conducting the statistical analysis, we have found the following:

Number of expression	Factors	Number
12	Thinking and problem solving	01
8	Patriotism and Identity	02
7	Planning and time management	03
7	Psychological and self-awareness	04
6	Language control	05
7	Scientific and technological	06
5	Communication	07
5	Social and working with the group	08

Table 2.
Tool axes and number of questions.

We notice a deviation in the distribution of scores from the normal distribution (multivariate distribution) in the parallel table. The Maximum Likelihood (ML) method retains its accuracy (parameter estimation) in the presence of a moderate level of deviation in the distribution of scores from the normal distribution (multivariate distribution) (**Table 3**)."

Variable	Min	Max	Skew	c.r.	Kurtosis	c.r.
Q57	1000	5000	−1602	−14,490	2547	11,522
Q50	1000	5000	−1501	−13,577	2360	10,673
Q51	1000	5000	−2182	−19,736	4449	20,124
Q52	1000	5000	−1980	−17,913	3572	16,155
Q53	1000	5000	−1650	−14,930	2310	10,448
Q54	1000	5000	−2390	−21,618	5835	26,393
Q55	1000	5000	−2675	−24,202	7733	34,978
Q56	1000	5000	−1089	−9854	,983	4447
Q42	1000	5000	−,417	−3772	−,767	−3471
Q37	1000	5000	−,786	−7113	,007	,030
Q38	1000	5000	−,467	−4224	−,601	−2719
Q39	1000	5000	−,554	−5010	−,240	−1087
Q40	1000	5000	−1546	−13,988	1853	8383
Q41	1000	5000	−,860	−7779	,192	,867
Q43	1000	5000	−,454	−4108	−,167	−,755
Q44	1000	5000	−,725	−6561	,068	,308
Q45	1000	5000	−,881	−7973	,240	1084
Q46	1000	5000	−,995	−8998	,318	1436
Q47	1000	5000	−,249	−2250	−,410	−1855
Q48	1000	5000	−1046	−9467	,759	3434
Q49	1000	5000	−1250	−11,303	1248	5643
Q8	1000	5000	−1337	−12,097	2008	9083
Q9	1000	5000	−,758	−6861	−,124	−,562
Q10	1000	5000	−,180	−1631	−,179	−,808
Q11	1000	5000	−,604	−5466	−,198	−,896
Q12	2000	5000	−,654	−5914	−,340	−1537
Q13	1000	5000	−,439	−3971	−,252	−1138
Q14	1000	5000	−,732	−6621	−,048	−,218
Q15	2000	5000	−,298	−2699	−,709	−3208
Q16	1000	5000	−,658	−5952	−,002	−,011
Q17	1000	5000	−,800	−7240	,450	2037
Q5	1000	5000	−,179	−1623	−,420	−1898
Q1	1000	5000	,025	,226	−,535	−2419

Variable	Min	Max	Skew	c.r.	Kurtosis	c.r.
Q2	1000	5000	−,501	−4534	−,394	−1784
Q3	1000	5000	−,740	−6698	−,091	−,413
Q4	1000	5000	−,938	−8482	,234	1057
Q6	1000	5000	−,068	−,614	−,509	−2301
Q7	1000	5000	−,915	−8273	,654	2957
Q18	1000	5000	−,154	−1393	−,683	−3091
Q19	1000	5000	−,714	−6459	,027	,124
Q20	1000	5000	−,872	−7889	,090	,406
Q21	1000	5000	−,386	−3489	−,664	−3006
Q22	1000	5000	−,383	−3464	−,676	−3060
Q23	1000	5000	−,319	−2882	−,632	−2861
Q24	1000	5000	−,427	−3865	−,681	−3079
Q25	1000	5000	−,162	−1468	−,124	−,563
Q26	1000	5000	−,097	−,880	−,612	−2768
Q36	1000	5000	−,377	−3414	−,542	−2452
Q35	1000	5000	−,182	−1643	−,336	−1520
Q27	1000	5000	,044	,396	−,618	−2794
Q32	1000	5000	−,117	−1055	−,859	−3884
Q28	1000	5000	−,265	−2400	−,630	−2849
Q29	1000	5000	−,027	−,243	−,692	−3132
Q30	2000	5000	−,289	−2610	−,814	−3680
Q31	1000	5000	−,050	−,448	−,735	−3323
Q33	1000	5000	−,318	−2874	−,541	−2445
Q34	1000	5000	−,684	−6186	−,101	−,459
Multivariate					**389,522**	**52,622**

Results of factor analysis SPSS version 24 assessment of normality.

Table 3.
Assessment of normality (Group number 1).

The ability of the model to estimate its parameters, or the designation of the model Identification Model:

It comes after the stage of building the model and determining it by the modeling method (**Table 4** and **Figure 1**).

Number of distinct sample moments:	1653
Number of distinct parameters to be estimated:	122
Degrees of freedom (1653–122)	1531

Confirmatory factor analysis Amos version 23.

Table 4.
Computation of degrees of freedom (Default model).

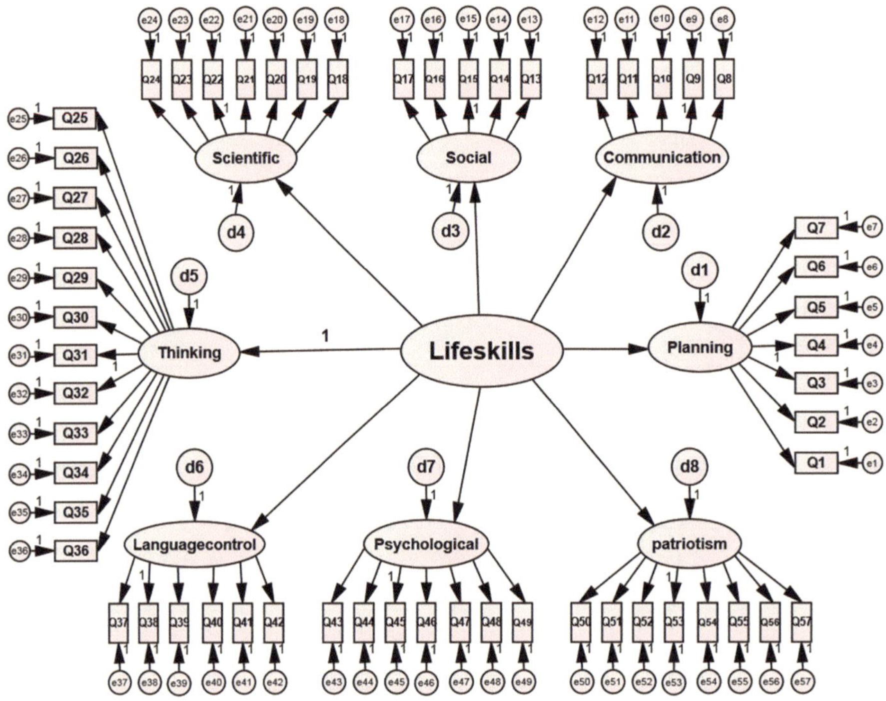

Figure 1.
Computation of degrees of freedom (Default model) [25].

Through the table and through a purely mathematical expression, we can express the first number 1653 as the units of information in the data:

Units of information in the data * 57 × (57 + 1) /2 = 1653 Sub-subsections can also be used throughout the manuscript.

While the number 125 expresses the needs of the model or unrestricted free parameters that need to be estimated:

1. the number of variations of independent variables, whether measured or latent: 66 (independent variables in this model are latent variables)

2. the number of correlations or variations between factors, latent variables or measurement errors (number of deleted arrows): 0 we mean by this the model from which we will initially proceed and which, through the results that will be sorted at its level, enables us to modify this model.

3. number of tracks (straight and unidirectional arrows) unrestricted: 56 Thus the needs of the model are: (66 + 0 + 56 = 122). Degree of freedom" (1653–122 = 1531).

Therefore, the model is specific and therefore its parameters can be estimated. Secondly, the study model's outputs and quality indicators:

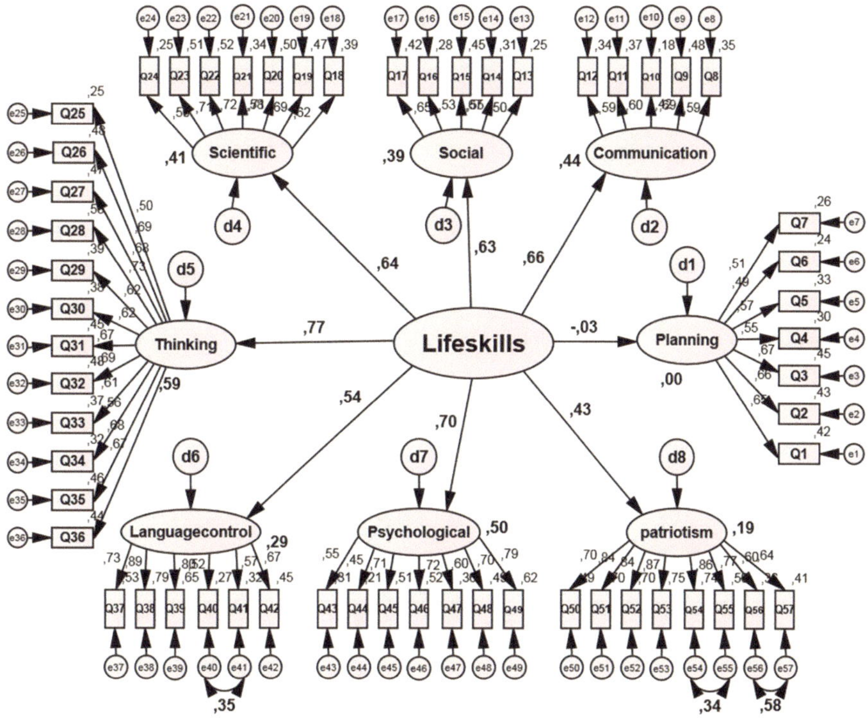

Figure 2.
Planning factor on life skills.

A. Estimation of the model's free parameters:

In this stage, we will attempt to estimate the standardized and non-standardized regression coefficients, covariances, and variations for the study model, which is essentially a measurement model. We will rely on the maximum likelihood (ML) method.

- The initial confirmatory model shows all the factors involved in the study:

See (**Figure 2**).

The confirmatory prototype shows all the factors involved in the study:

Confirmatory factor analysis Amos version 23.

By the graph of the modular solution of the model under study, and by observing the standard regression coefficients, it is clear that all questions.

The saturation of the axes that you interpret has exceeded 40%, What is striking is the lack of saturation of the planning factor on life skills in the form sufficient -, 032 where there is almost no amount of information that is interpreted in this The axis.

3.2.2.4 Unstandardized regression coefficients or raw scores

See (**Table 5**).

			Estimate	S.E.	C.R.	P
Planning	<---	Lifeskills	-,043	,076	–,567	,571
Communication	<---	Lifeskills	,741	,086	8668	***
patriotism	<---	Lifeskills	,643	,087	7375	***
Social	<---	Lifeskills	,705	,085	8265	***
Psychological	<---	Lifeskills	,910	,096	9494	***
Scientific	<---	Lifeskills	,865	,095	9071	***
Thinking	<---	Lifeskills	1000			
Languagecontrol	<---	Lifeskills	,866	,099	8735	***
Q31	<---	Thinking	,915	,066	13,760	***
Q21	<---	Scientific	,809	,068	11,868	***
Q24	<---	Scientific	,736	,072	10,228	***
Q19	<---	Scientific	,864	,062	13,871	***
Q22	<---	Scientific	1000			
Q23	<---	Scientific	,909	,063	14,385	***
Q18	<---	Scientific	,802	,063	12,651	***
Q20	<---	Scientific	,854	,060	14,235	***
Q29	<---	Thinking	,776	,061	12,821	***
Q30	<---	Thinking	,782	,061	12,771	***
Q32	<---	Thinking	1000			
Q28	<---	Thinking	,942	,063	14,889	***
Q27	<---	Thinking	,868	,062	13,976	***
Q34	<---	Thinking	,679	,058	11,619	***
Q35	<---	Thinking	,901	,065	13,879	***
Q26	<---	Thinking	,832	,059	14,175	***
Q25	<---	Thinking	,619	,059	10,465	***
Q36	<---	Thinking	,883	,065	13,668	***
Q33	<---	Thinking	,842	,067	12,587	***
Q4	<---	Planning	,757	,075	10,070	***
Q2	<---	Planning	,976	,084	11,566	***
Q3	<---	Planning	1000			
Q5	<---	Planning	,850	,082	10,408	***
Q1	<---	Planning	,846	,074	11,459	***
Q7	<---	Planning	,551	,058	9424	***
Q6	<---	Planning	,696	,077	9077	***
Q15	<---	Social	1000			
Q13	<---	Social	,784	,088	8957	***
Q16	<---	Social	,869	,094	9283	***
Q17	<---	Social	,992	,092	10,790	***

			Estimate	S.E.	C.R.	P
Q14	<---	Social	,863	,089	9662	***
Q10	<---	Communication	,669	,086	7758	***
Q8	<---	Communication	,823	,079	10,409	***
Q11	<---	Communication	,934	,088	10,567	***
Q12	<---	Communication	,798	,077	10,324	***
Q9	<---	Communication	1000			
Q46	<---	Psychological	,959	,066	14,617	***
Q49	<---	Psychological	,986	,062	15,786	***
Q44	<---	Psychological	,637	,069	9297	***
Q47	<---	Psychological	,748	,061	12,169	***
Q48	<---	Psychological	,854	,060	14,217	***
Q43	<---	Psychological	,649	,057	11,294	***
Q45	<---	Psychological	1000			
Q39	<---	Languagecontrol	,963	,045	21,250	***
Q37	<---	Languagecontrol	,736	,040	18,485	***
Q40	<---	Languagecontrol	,581	,048	11,988	***
Q41	<---	Languagecontrol	,631	,048	13,233	***
Q38	<---	Languagecontrol	1000			
Q42	<---	Languagecontrol	,803	,049	16,369	***
Q53	<---	Patriotism	1000			
Q56	<---	Patriotism	,668	,045	14,699	***
Q51	<---	Patriotism	,882	,037	23,897	***
Q54	<---	Patriotism	,871	,035	25,004	***
Q55	<---	Patriotism	,719	,035	20,469	***
Q50	<---	Patriotism	,727	,040	18,139	***
Q52	<---	Patriotism	,906	,038	23,891	***
Q57	<---	Patriotism	,689	,043	16,024	***

Confirmatory factor analysis Amos version 23.

Table 5.
Regression weights: (Group number 1 - Default model).

By the table of non-standard regression coefficients, i.e. saturation of questions on the axes that you interpret all came up as a function statistically (P is less than5%), and it is striking the lack of saturation of the planning factor on life skills.

Where his non-standard regression coefficient came non-statistically d,5710 (P greater than 5%) (**Table 6**).

Through both tables, standardized and unstandardized regression coefficients, and the graphical representation, it is evident that all the questions are significantly associated with the axes they are meant to explain, with substantial saturations exceeding

			Estimate
Planning	<---	Lifeskills	-,032
Communication	<---	Lifeskills	,662
patriotism	<---	Lifeskills	,431
Social	<---	Lifeskills	,625
Psychological	<---	Lifeskills	,704
Scientific	<---	Lifeskills	,644
Thinking	<---	Lifeskills	,766
Languagecontrol	<---	Lifeskills	,542
Q31	<---	Thinking	,671
Q21	<---	Scientific	,584
Q24	<---	Scientific	,502
Q19	<---	Scientific	,686
Q22	<---	Scientific	,723
Q23	<---	Scientific	,713
Q18	<---	Scientific	,623
Q20	<---	Scientific	,705
Q29	<---	Thinking	,623
Q30	<---	Thinking	,620
Q32	<---	Thinking	,693
Q28	<---	Thinking	,730
Q27	<---	Thinking	,682
Q34	<---	Thinking	,562
Q35	<---	Thinking	,677
Q26	<---	Thinking	,693
Q25	<---	Thinking	,504
Q36	<---	Thinking	,666
Q33	<---	Thinking	,611
Q4	<---	Planning	,551
Q2	<---	Planning	,656
Q3	<---	Planning	,668
Q5	<---	Planning	,574
Q1	<---	Planning	,648
Q7	<---	Planning	,510
Q6	<---	Planning	,489
Q15	<---	Social	,674
Q13	<---	Social	,504
Q16	<---	Social	,527
Q17	<---	Social	,647

			Estimate
Q14	<---	Social	,554
Q10	<---	Communication	,419
Q8	<---	Communication	,593
Q11	<---	Communication	,605
Q12	<---	Communication	,586
Q9	<---	Communication	,690
Q46	<---	Psychological	,724
Q49	<---	Psychological	,789
Q44	<---	Psychological	,454
Q47	<---	Psychological	,597
Q48	<---	Psychological	,703
Q43	<---	Psychological	,553
Q45	<---	Psychological	,713
Q39	<---	Languagecontrol	,803
Q37	<---	Languagecontrol	,729
Q40	<---	Languagecontrol	,523
Q41	<---	Languagecontrol	,567
Q38	<---	Languagecontrol	,889
Q42	<---	Languagecontrol	,667
Q53	<---	Patriotism	,868
Q56	<---	Patriotism	,604
Q51	<---	Patriotism	,836
Q54	<---	Patriotism	,859
Q55	<---	Patriotism	,766
Q50	<---	Patriotism	,703
Q52	<---	Patriotism	,836
Q57	<---	Patriotism	,644

Confirmatory factor analysis Amos version 23.

Table 6.
Standardized regression weights: (Group number 1 - Default mode).

40%. In this regard, all unstandardized regression coefficients are statistically significant (P < 5%). Notably, it's interesting to observe the lack of saturation of the planning factor on life skills, with a coefficient of −0.032. This implies that there is minimal information explained by this factor. The unstandardized regression coefficient for this factor is also not statistically significant at 0.5710 (P > 5%).

3.2.2.5 The variations (common variance) and covariances

See (**Tables** 7 and **8**).

			Estimate	S.E.	C.R.	P	Label
e56	<-->	e57	,263	,025	10,382	***	
e41	<-->	e40	,220	,032	6851	***	
e55	<-->	e54	,068	,012	5633	***	

Confirmatory factor analysis Amos version 23.

Table 7.
Covariances: (Group number 1 - Default model).

			Estimate
e56	<-->	e57	,575
e41	<-->	e40	,350
e55	<-->	e54	,342

Confirmatory factor analysis Amos version 23.

Table 8.
Correlations: (Group number 1 - Default model).

These variations and Covariances in the two tables were obtained after modifying the model. This was done using adjustment indicators to achieve the maximum convergence or match between the model and the data. This results in a better interpretation of information by the factors, leading to an overall improvement in the quality of the fit.

Model testing, or testing the quality of model fit:

Through various types of fit indices, we obtain a general or overall assessment of how well the model matches the data. We will review the results of widely used fit indices together.

It is carried out through conformity indicators of various types and they provide us with a general or aggregate about the conformity of the model to the data, and we will review together the results of widely used or used conformity indicators (**Table 9**).

We notice from the table that the first line is specific to our model, in the second line which is the saturated model while the last line is the independent model independent Model or the null model, which is based on the assumption that the variances of the observed variables on the level of society is equal to zero or zero and only the values of the variance of these variables remain (the model with independent variables).

Degree of Freedom = 1528, the value was 2499,040 = CMIN, which is a function statistically, 1635 = CMIN/DF, which is less than 3, it is good for these two indicators within the matching indicators absolute fit indices and indicate to what extent the

Model	NPAR	CMIN	DF	P	CMIN/DF
Default model	125	2499,040	1528	,000	1635
Saturated model	1653	,000	0		
Independence model	57	12,911,951	1596	,000	8090

Confirmatory factor analysis Amos version 23.

Table 9.
CMIN.

Model	RMR	GFI	AGFI	PGFI
Default model	,041	,850	,837	,785
Saturated model	,000	1000		
Independence model	,176	,272	,246	,263
Confirmatory factor analysis Amos version 23.				

Table 10.
RMR, GFI.

information derived from the model is represented Presumably the information contained in the data of the research sample (**Table 10**).

We observe from the table that:

The value,850 = GFI, which is inappropriate is less than 0.90 within the absolute conformity indicators.

Absolute fit indices (**Table 11**).

We note through the table that the value of, 910 = TLI and the value of, 910 = CFI which are acceptable values for the two indices are bounded between 0.90 and 0.95 within the comparative or incremental Comparative fit matching indicators indices/fit indices incremental (**Table 12**).

We note from the table that: the value of ,036 = RMSEA, which is a good indicator among the indicators of correcting the lack of economy or economic indicators correction indices parsimony (**Table 13**).

We note from the table that:

The economic indicators AICO BCC and BIC of the study model reached lower values compared to the saturated model They are a good indicator of the economy; these indicators are among the indicators of correcting the lack of economy or economic indicators correction indices parsimony (**Table 14**).

Through the table.

Model	NFI Delta1	RFI rho1	IFI Delta2	TLI rho2	CFI
Default model	,806	,798	,915	,910	,914
Saturated model	1000		1000		1000
Independence model	,000	,000	,000	,000	,000
Confirmatory factor analysis Amos version 23.					

Table 11.
Baseline comparisons.

Model	RMSEA	LO 90	HI 90	PCLOSE
Default model	,036	,033	,039	1000
Independence model	,120	,118	,122	,000
Confirmatory factor analysis Amos version 23.				

Table 12.
RMSEA.

Model	AIC	BCC	BIC	CAIC
Default model	2749,040	2782,605	3273,596	3398,596
Saturated model	3306,000	3749,861	10,242,722	11,895,722
Independence model	13,025,951	13,041,257	13,265,149	13,322,149

Confirmatory factor analysis Amos version 23.

Table 13.
AIC.

Model	ECVI	LO 90	HI 90	MECVI
Default model	5610	5339	5898	5679
Saturated model	6747	6747	6747	7653
Independence model	26,584	25,851	27,329	26,615

Confirmatory factor analysis Amos version 23.

Table 14.
ECVI.

The ECVI economic index of the study model reached a lower value compared to the saturated model and the independent model, namely A good indicator of the economy, these indicators are among the indicators of correcting the lack of economy or economic indicators correction indices parsimony.

SRMR = 0.048.

The value of SRMR = 0.048 was a good indicator for the economy among the indicators for correcting the lack of economy or economic indicators correction indices parsimony (**Table 15**).

Based on the matching indicators results shown in the table, it appears that they are appropriate. This is confirmed by the results obtained from the confirmatory factor analysis, where the model demonstrated a good fit for the study sample data.

Bollen-Stine Bootstrap (Default model)

The model fit better in 9999 bootstrap samples.

It fit about equally well in 0 bootstrap samples.

Match Indicators	Standard	Value	Judgment
CMIN (χ^2)	Non-significant	2060.40	Non-significant (p = 0.00)
D_f	Degree of freedom	1165	
N_c	Good (1–3) (3–5) Acceptable	1.76	Good
SRMR	Good (0–0.05) (0.05–0.08) Acceptable	0.05	Good
GFI (Goodness of Fit Index)	0.9–0.95, (Acceptable) 0.95–1 (Good)	.86	Not suitable
TLI (Tucker-Lewis Index)		0.91	Acceptable
CFI (Comparative Fit Index)		0.92	
CFI (Comparative Fit Index)		0.92	

Match Indicators	Standard	Value	Judgment
RMSEA (Root Mean Square Error of Approximation)	Good (0–0.05) (0.05–0.08) Acceptable	0.04	Good

Confirmatory factor analysis Amos version 23.

Table 15.
A Fit indices for the initial model after modification.

It fit worse or failed to fit in 1 bootstrap samples.

Testing the null hypothesis that the model is correct, Bollen-Stine bootstrap p =,000.

The same results that we obtained previously indicate that the number of samples used for bootstrapping the matching indicators was 9999 out of 10,000 samples, with only one sample showing an inappropriate model-data fit.

Hypothesis Testing: Based on these results, the statistical difference is significant ($p = 0.000$), which is less than 0.05 (the significance level). Therefore, we reject the null hypothesis, indicating statistically significant differences in favor of the model's quality and suitability. It demonstrated good data fit for the vast majority.

From the results obtained through the confirmatory factor analysis of the proposed model for estimating free parameters and the goodness of fit test, it is evident that the model is suitable for measuring life skills, pending the comparison between the two models.

4. The ability of the model to estimate its parameters, or the designation of the model identification model

It comes after the stage of building the model and determining it by the modeling method Units of information in the data * $50 \times (50 + 1)/2 = 1275$.

The number 107 expresses the needs of the model or unrestricted free parameters that need to be estimated: 1-the number of variations of independent variables, whether measured or latent: 58 (independent variables in this model are latent variables) 2-the number of correlations or variations between factors, latent variables or measurement errors (number of deleted arrows): 0 3. number of tracks (straight and unidirectional arrows) unrestricted: 49 Thus the needs of the model are: (58+ 0 + 49 = 107).

Degree of freedom" (1275–107 = 1168) Therefore, the model is specific and therefore its parameters can be estimated.

It is worth mentioning and what we should draw attention to is that the appointment of the model is before the researcher goes into collecting the results, otherwise he will get into trouble, he may finish collecting the results of the research, which will cost time, and then he comes to test his model, which he can find is not assigned (in case of absence of appointment) (**Figure 3**).

The study model after the exclusion of the planning axis.

Confirmatory factor analysis Amos version 23.

Computation of degrees of freedom (Default model).

Number of distinct sample moments: 1275

Number of distinct parameters to be estimated: 107

Degrees of freedom (1275–107): 1168

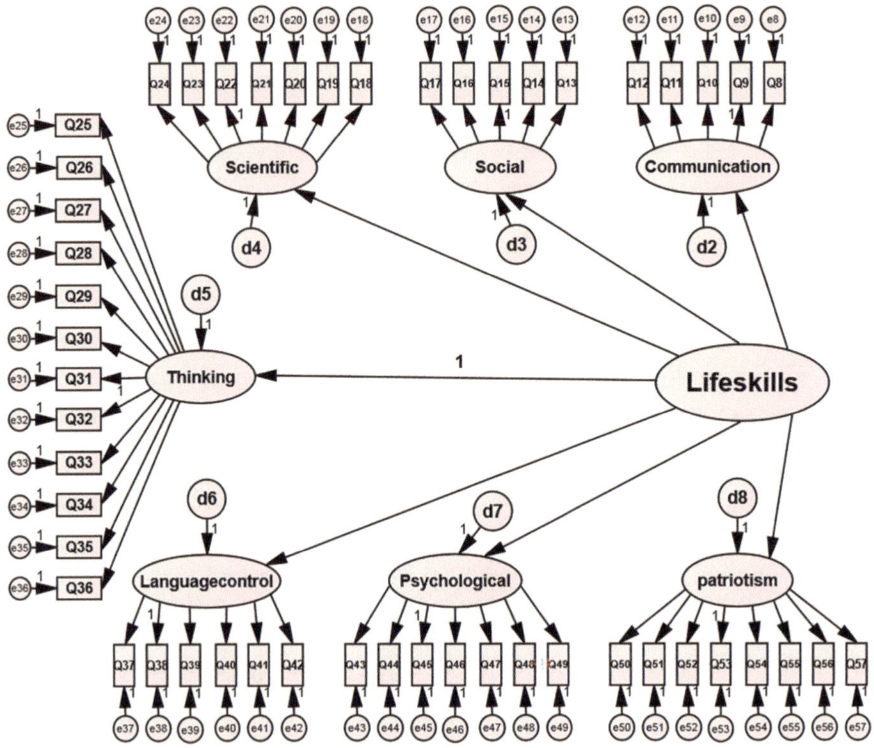

Figure 3.
A-modeling estimation of free parameters (Parameter estimation).

4.1 Outputs and indicators of the quality of the study model

4.1.1 A-modeling estimation of free parameters (Parameter estimation)

At this stage, we will try to estimate: standard and non-standard regression coefficients, correlations and variations of the study model, which is actually a measurement model, where we will rely on the maximum probability method (ML).

4.1.1.1 Outputs and indicators

Modified confirmatory factor model after exclusion of planning (**Figure 4**).

Confirmatory factor analysis Amos version 23.

By the graph of the modular solution of the model without the diagram under consideration, and by observing the standard regression coefficients, clearly shows that all questions are saturated on the interlocutor that you interpret saturations.

Considering that she exceeded 40%, which is the same for all axes on her parent worker life skills above 40%.

Standard and non-standard regression coefficients (**Table 16**).

Through the table the non-standard regression coefficients i.e. the saturations the rough grades of the questions on the axes that you interpret came all of them are

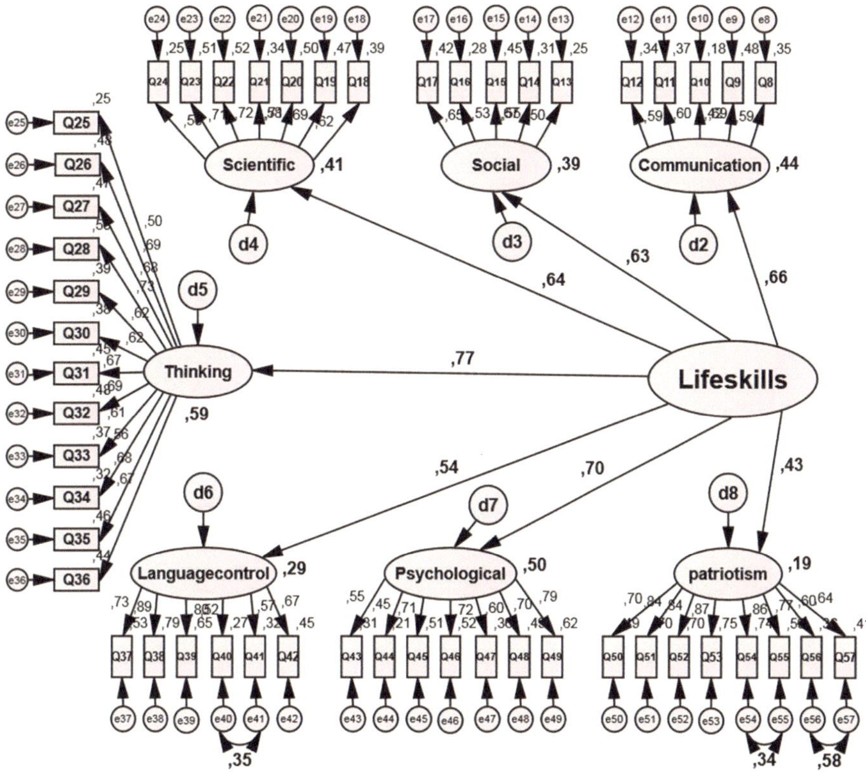

Figure 4.
Standardized regression weights.

			Estimate	S.E.	C.R.	P
Social	<---	Life skills	,705	,085	8268	***
Psychological	<---	Life skills	,910	,096	9497	***
Scientific	<---	Life skills	,864	,095	9067	***
Thinking	<---	Life skills	1000			
Language control	<---	Life skills	,865	,099	8732	***
Communication	<---	Life skills	,741	,086	8668	***
Patriotism	<---	Life skills	,643	,087	7375	***
Q31	<---	Thinking	,915	,066	13,760	***
Q21	<---	Scientific	,809	,068	11,868	***
Q24	<---	Scientific	,736	,072	10,226	***
Q19	<---	Scientific	,864	,062	13,870	***
Q22	<---	Scientific	1000			
Q23	<---	Scientific	,909	,063	14,387	***
Q18	<---	Scientific	,802	,063	12,652	***
Q20	<---	Scientific	,854	,060	14,234	***
Q29	<---	Thinking	,776	,061	12,822	***

			Estimate	S.E.	C.R.	P
Q30	<---	Thinking	,781	,061	12,770	***
Q32	<---	Thinking	1000			
Q28	<---	Thinking	,942	,063	14,891	***
Q27	<---	Thinking	,868	,062	13,977	***
Q34	<---	Thinking	,679	,058	11,619	***
Q35	<---	Thinking	,901	,065	13,878	***
Q26	<---	Thinking	,832	,059	14,175	***
Q25	<---	Thinking	,619	,059	10,466	***
Q36	<---	Thinking	,883	,065	13,668	***
Q33	<---	Thinking	,842	,067	12,586	***
Q15	<---	Social	1000			
Q13	<---	Social	,784	,088	8957	***
Q16	<---	Social	,869	,094	9285	***
Q17	<---	Social	,992	,092	10,792	***
Q14	<---	Social	,863	,089	9661	***
Q10	<---	Communication	,670	,086	7758	***
Q8	<---	Communication	,823	,079	10,409	***
Q11	<---	Communication	,934	,088	10,565	***
Q12	<---	Communication	,798	,077	10,325	***
Q9	<---	Communication	1000			
Q46	<---	Psychological	,959	,066	14,616	***
Q49	<---	Psychological	,986	,062	15,784	***
Q44	<---	Psychological	,637	,069	9299	***
Q47	<---	Psychological	,748	,061	12,171	***
Q48	<---	Psychological	,854	,060	14,216	***
Q43	<---	Psychological	,649	,057	11,295	***
Q45	<---	Psychological	1000			
Q39	<---	Language control	,962	,045	21,249	***
Q37	<---	Language control	,736	,040	18,486	***
Q40	<---	Language control	,581	,048	11,988	***
Q41	<---	Language control	,630	,048	13,232	***
Q38	<---	Language control	1000			
Q42	<---	Language control	,803	,049	16,369	***
Q53	<---	patriotism	1000			
Q56	<---	patriotism	,668	,045	14,699	***
Q51	<---	patriotism	,882	,037	23,897	***
Q54	<---	patriotism	,871	,035	25,003	***
Q55	<---	patriotism	,719	,035	20,469	***

			Estimate	S.E.	C.R.	P
Q50	<---	patriotism	,727	,040	18,139	***
Q52	<---	patriotism	,906	,038	23,891	***
Q57	<---	patriotism	,689	,043	16,024	***

Confirmatory factor analysis Amos version 23.

Table 16.
Regression weights: (Group number 1 - Default model).

statistically a function (P is less than 5%), the same for all regressions that form between the axes involved in the study on the worker the mother or the basic axis (life skills) where the other regression coefficient her standard came statistically D (P less than 5%).

4.1.1.2 Standardized regression weights

See (**Table 17**).

Through the tables, the standard and non-standard regression coefficients and the graph clearly show that all the questions were saturated on the axes interpreted by significant saturations that exceeded 40%, so that the non- standard regression coefficients all came statistically (P less than 5%), we also note the saturations of factors or axes on their basic factor (life skills) all came at the level where they all exceeded 0.40.

Where all of its non-standard regression coefficients came up as a function statistically (P less than 5%).

			Estimate
Social	<---	Life skills	,625
Psychological	<---	Life skills	,704
Scientific	<---	Life skills	,643
Thinking	<---	Life skills	,767
Language control	<---	Life skills	,542
Communication	<---	Life skills	,662
patriotism	<---	Life skills	,431
Q31	<---	Thinking	,671
Q21	<---	Scientific	,584
Q24	<---	Scientific	,502
Q19	<---	Scientific	,686
Q22	<---	Scientific	,723
Q23	<---	Scientific	,713
Q18	<---	Scientific	,623
Q20	<---	Scientific	,705
Q29	<---	Thinking	,623

			Estimate
Q30	<---	Thinking	,620
Q32	<---	Thinking	,693
Q28	<---	Thinking	,730
Q27	<---	Thinking	,682
Q34	<---	Thinking	,562
Q35	<---	Thinking	,677
Q26	<---	Thinking	,693
Q25	<---	Thinking	,504
Q36	<---	Thinking	,666
Q33	<---	Thinking	,611
Q15	<---	Social	,674
Q13	<---	Social	,504
Q16	<---	Social	,527
Q17	<---	Social	,647
Q14	<---	Social	,554
Q10	<---	Communication	,419
Q8	<---	Communication	,593
Q11	<---	Communication	,604
Q12	<---	Communication	,586
Q9	<---	Communication	,690
Q46	<---	Psychological	,724
Q49	<---	Psychological	,789
Q44	<---	Psychological	,454
Q47	<---	Psychological	,598
Q48	<---	Psychological	,703
Q43	<---	Psychological	,553
Q45	<---	Psychological	,713
Q39	<---	Language control	,803
Q37	<---	Language control	,729
Q40	<---	Language control	,523
Q41	<---	Language control	,567
Q38	<---	Language control	,889
Q42	<---	Language control	,667
Q53	<---	Patriotism	,868
Q56	<---	Patriotism	,604
Q51	<---	Patriotism	,836
Q54	<---	Patriotism	,859
Q55	<---	Patriotism	,766

			Estimate
Q50	<---	Patriotism	,703
Q52	<---	Patriotism	,836
Q57	<---	Patriotism	,644
Confirmatory factor analysis Amos version 23.			

Table 17.
Standardized regression weights:(Groupnumber1-efaultmodel).

4.1.1.2.1 Covariances (covariance) and correlations

See (**Table 18**).

These variances by the table were obtained after modifying the model and this is through the modification indices modification indicators, which is a function statistically (P is less than 5%) this is in order to achieve maximum convergence or congruence between the model and the data, which leads to a better interpretation of the information by the factors thus improve the quality of the match in general (**Table 19**).

The correlations in the table as well were obtained after modifying the model and this through the indicators of modification indices modification where it was the correlations of these errors in the level leading to a better explanation for information by factors to improve correlations among themselves, which leads to improve the quality of matching (**Table 20**).

We notice from the table that the first line is for our no-planning study form in the second line is the ideal saturated model while the last line is the independent model or null model null model which is based on the assumption that the variances of variables.

The observation at the community level is equal to zero or zero and only the values of the variance of these variables remain (The model with independent variables).

	Estimate S.E.C.R.P Label
e56–e57	,263,02510,382***
e41–e40	,220,0326,852***
e55–e54	,068,0125,633***
Confirmatory factor analysis Amos version 23.	

Table 18.
Covariances: (Groupnumber1-Defaultmodel).

	Estimate
e56–e57	,575
e41–e40	,350
e55–e54	,342
Confirmatory factor analysis Amos version 23.	

Table 19.
Correlations:(Groupnumber1-Defaultmodel).

Model	NPAR	CMIN	DF	P	CMIN/DF
Default model	110	2060,401	1165	,000	1769
Saturated model	1275	,000	0		
Independence model	50	11,762,350	1225	,000	9602
Confirmatory factor analysis Amos version 23.					

Table 20.
CMIN.

Degree of Freedom = 1165, the value was 2060,401 = CMIN, which is a function statistically.

1769 = CMIN/DF, which is less than 3, so these two indicators are good within the matching indicators absolute fit indices and indicate to what extent the information derived from the model is represented presumably the information contained in the data of the research sample (**Table 21**).

We note through a table within the outputs of the study model without a planning axis.

The value ,850 = GFI, which is inappropriate is less than 0.90 within the absolute conformity indicators.

Absolute fit indices.

The ability of the model to estimate its parameters, or the designation of the model Identification Model (**Table 22**).

We note through the table within the outputs of the study model without the planning axis reached a value of, 910 = TLI and reached a value of, 910 = CFI, which are acceptable values for indicators limited between0.90 and 0.95 within the comparative or incremental Comparative fit matching indicators indices/fit indices incremental (**Table 23**).

Model	RMR	GFI	AGFI	PGFI
Default model	,041	,856	,842	,782
Saturated model	,000	1000		
Independence model	,195	,255	,225	,245
Confirmatory factor analysis Amos version 23.				

Table 21.
RMR, GFI.

Model	NFI delta1	RFI rho1	IFI delta2	TLI rho2	CFI
Default model	,825,816	,916	,911	,915	
Saturated model	1000		1000		1000
Independence model	,000	,000	,000	,000	,000
Confirmatory factor analysis Amos version 23.					

Table 22.
Baseline comparisons.

Model	RMSEA	LO 90	HI 90	PCLOSE
Default model	,040	,037	,042	1000
Independence model	,132	,130	,135	,000

Confirmatory factor analysis Amos version 23.

Table 23.
RMSEA.

Through the table: within the outputs of the study model without the planning axis a value of ,040 = RMSEA, which is a good indicator among the indicators of correcting the lack of economy or economic indicators correction indices parsimony (**Table 24**).

Through a table within the outputs of the study model without a planning axis the economic indicators AICO BCC and BIC of the study model without planning reached lower values compared to the saturated model is a good indicator of the economy, these indicators are among the indicators of correcting the lack of economy or economic indicators correction indices parsimony (**Table 25**).

Through a table within: the outputs of the study model without a planning axis.

The ECVI economic index of the study model without planning reached a lower value compared to the saturated model and the model the Independent is a good indicator of the economy, these indicators are among the indicators of correcting the lack of economy or economic indicators correction indices parsimony.

SRMR value was = 0.05, which is a good indicator for the economy among the indicators for correcting the lack of economy or economic indicators correction indices parsimony.

Model	AIC	BCC	BIC	CAIC
Default model	2280,401	2305,959	2742,010	2852,010
Saturated model	2550,000	2846,241	7900,466	9175,466
Independence model	11,862,350	11,873,967	12,072,172	12,122,172

Confirmatory factor analysis Amos version 23.

Table 24.
AIC.

Model	ECVI	LO 90	HI 90	MECVI
Default model	4654	4404	4920	4706
Saturated model	5204	5204	5204	5809
Independence model	24,209	23,507	24,924	24,233

Confirmatory factor analysis Amos version 23.

Table 25.
ECVI.

Conformity Indicators	The standard	Value	Judgment
$CMIN_x2$	Not signified	2060,40	Signified p = 0.00
D f	Degree of freedom	1165	
N c	1–3 Good 3–5 Acceptable	1,76	Good
SRMR	0–0.05 Good 0.08–0.05 Acceptable	0.05	Good
GFI	0.9–0.95 Acceptable	,86	Not suitable
TLI	0.95–1 Good	0.91	Acceptable
CFI	0.95–1 Good	0.92	Acceptable
RMSEA	0–0.05 Good 0.05–0.08 Acceptable	0.04	Good

Confirmatory factor analysis Amos version 23.

Table 26.
ECVI.

Table represent conformity indicators in favor of the study model after the exclusion of planning and after adjustment (**Table 26**).

Through the results of the conformity indicators shown in the table, it appears that they are appropriate, and this is confirmed by these results obtained by confirmatory factor analysis, where the model showed a good match for the data of the study sample.

The same results as we got earlier: according to which the number of Bootstrap samples in which the matching indicators worked or were good at their level is 9999 out of 10,000 samples, while one sample in which the model matching of the data was inappropriate.

Testing the null hypothesis: through these results, the difference is statistically D p =, 000, which is less than 0.05 (the level of significance), and therefore we reject the null hypothesis, and therefore there are significant differences in favor of the quality and validity of the model, as it showed quality in matching the data at the overwhelming majority.

Through the results we have reached through the confirmatory factor analysis of the proposed model in estimating the modeling of free parameters, as well as testing the quality of the model's data matching, it is evident that it is suitable for measuring life skills, waiting for the results of the comparison between the two models.

4.1.1.2.2 Make a comparison between the two models

The initial study model that includes all the factors used in factor analysis a c p, the study model without a factor or planning axis (**Table 27**).

Conformity indicators	The standard	Model (1) study without factor or planning axis		Model (2) study without factor or planning axis	
		Value	Judgment	value	judgment
CMIN χ2	Not signified	2499.04	P = 0.00 signified	2060.4	P = 0.00 signified
D f	Degree of freedom	1528		1165	
AIC	Used to compare models: the model with the smallest value is the best	2749		2280	Matching the second form to the sample data is better than matching the first form
BCC		2782		2305	
BIC		3273		2742	
ECVI		5.61		4.65	
Confirmatory factor analysis Amos version 23.					

Table 27.
The study model without a factor or planning axis.

By comparing the first and second models using the economic conformity indicators that are used to compare the models, we conclude that the second model without the planning axis is the most appropriate to measure life skills because these conformity indicators are at its best, where they have the smallest value, and this confirms the results we obtained in the factor analysis, where the planning axis is not saturated with the rest of the factors within the factor that combines them.

Through all the above, we can point out that the model using confirmatory factor analysis came to confirm what we get in exploratory factor analysis.

Therefore, we can adopt the model in measuring life skills among university students. As follows:

The ability of the model to estimate its parameters, or the designation of the model Identification Model (**Table 28**).

Factors
Thinking and problem solving
Have sound critical thinking
I have the ability to identify problems specific to my field of specialization
The ability to propose appropriate solutions to each problem.
I have the ability to organize thoughts in a logical way
I have the ability to find alternatives to the problem
I have the ability to think independently.
I have the ability to analyze.
I have the ability to research and experiment.
I can relate educational situations to similar life situations.
I can sense the problem.
I have the ability to accurately identify the problem.
I can collect information about the subject and its parent.

Density skills and patriotism and Identity Respect national symbols.
I defend and protect my homeland.
I work hard to serve my country.
I enjoy the love of the Fatherland.
I am proud to belong to my homeland.
Apply general rules and regulations
I respect the National Law.
I act with the credibility of a national trend.
Psychological skills and self-awareness Psychological and self-awareness
I can control the situations that confront me.
I am often proud of what I do.
I have high self-confidence.
I have the ability to adjust my feelings .
I have the ability to detect other people's feelings.
I predict the expected situations.
I have the ability of self-reliance.
Language control skills Language control
I understand the meanings of the English language,
I have the ability to discuss in a sound language and present research that is prepared by me in front of the professor and colleagues.
I can translate the basic terms of the specialty from Arabic to French.
I have the ability to express in proper Arabic.
I can intervene in discussions in proper Arabic.
I acquire the skill of translation into multiple languages.
Scientific and technological scientific and technological skills. I can use the computer skillfully.
I keep abreast of modern scientific and technical developments.
I have the ability to use modern technological means.
I adhere to the basics of scientific research and its ethics.
I use different sources to obtain information and knowledge in order to serve the cognitive outcome.
I acquire a diverse knowledge culture.
Gain the skill of using the internet.
Communication and communication //communication skills It's better to listen to others.
Improve verbal communication.
I express my thoughts clearly.
I use the appropriate vocabulary when talking to others.
Listen attentively to the words of others.
Social skills and working with the group Social and working with the group
I treat others on the basis of tolerance.
I accept and respect another point of view.

I live with other people's problems.
I have the ability to put the interest of the group over the interest of the individual.
I have the ability to build bonds of trust with others.

Table 28.
Parameters (Questions), or the designation of the model Identification Model.

5. Conclusion

The construction of metrics is one of the important things that have become used today in scientific research and through which credible results are obtained.

The construction of the life skills scale has passed through two important stages, the first stage was represented by the two exploratory factors and the second stage was represented by the confirmatory factor analysis, through which we reached the construction of two models using economic conformity indicators that are used to compare the models, where we concluded that the second model without the planning axis is the most appropriate to measure life skills because these conformity indicators in its efficiency where it has the smallest value.

From the foregoing, we can point out that the model using confirmatory factor analysis came to confirm what we get in exploratory factor analysis. Therefore, we can adopt the model in measuring the life skills of university students.

Abbreviations

SRMR	Standardized Root Mean squared Residual
DF	Degree of freedom
S.E	Standard Error human serum
CR	Critical Ratio
NPAR	Number of Parameters for each model
RMS	Root mean square residual
GFI	Goodness of Fit Index
AGFI	Adjusted Good ness of Fit Index

Author details

Atallah Ahmed[1*], Touati Hayat[2], Saad Mohammed Abdoulmoudjib[3], Amrani Amel[2], Berrabah Ameur[2], Cherifi Selma[4], Allali Taleb[5] and Benkhaled Hadj[5]

1 Laboratory Optimization of Sports Activity Programs4 LABOPAPS (CODE W0890400), Theory and Methodology of Physical Education, Abdul Hamid bin Badis University Mostaganem, Algeria

2 Abdelhamid Ben Badis University of Mostaganem, Algeria

3 Laboratory Optimization of Sports Activity Programs LABOPAPS (CODE W0890400), Measurement and Evaluation, Algeria

4 Theory and Methodology of Physical Education, Dely Ibrahim University of Alger 3, Algeria

5 Theory and Methodology of Physical Education, Abdelhamid Ben Badis University of Mostaganem, Algeria

*Address all correspondence to: ahmed.atallah@univ-mosta.dz

References

[1] Teo T. Pre-service teachers' attitudes towards computer use: A Singapore survey. Australasian Journal of Educational Technology. 2008;**24**:413-424

[2] Amna KA-H. The reality of life skills development: An analytical study of the content of Arabic language curricula in secondary education. Journal of the Union of Arab Universities for Education and Psychology. 2015;**13**(1):178-203

[3] Subhi Lulu F, Qishta AS. The Level of Life Skills among Students Graduating from the Faculty of Education at the Islamic University of Gaza. (2005-2006)

[4] Obeidat O, Suad S. Skills available in Jordanian higher education outputs as required by local work. Arab Journal for Quality Assurance of Higher Education. 2010;**III**(5)

[5] Hartmann A. Longitudinal study of the quality of life and adjustment strategies of breast cancer patients and their "companion-referent". Psychology. 2007. Thesis presented with a view to obtaining the degree of Doctor in Sychology 14 December 2007, University of Rennes 2, 2007. French. <tel-00267588>, HAL Id: tel-00267588. Available from: https://tel.archives-ouvertes.fr/tel-00267588 [Accessed:March 27, 2008]

[6] Magdy AKH. The effectiveness of using information technologies in achieving the dimensions of quality of life among samples of Omani students, In: Proceedings of the Psychology and Wellbeing Symposium; 18-19 December; Sultan Qaboos University. 2006

[7] Gatab TA, Shayanb N, Tazangic RM, Taherid M. Students' life quality prediction based on life skills, Sciverse Science Derect. Procidia –Social and Behavioral Sciences. 2011;**30**:1980-1982. DOI: 10.1016/j.sbspro.2011.10.384

[8] Omar A. Contributions of some modern teaching methods in physical education and sports to the development of some life skills among undergraduate students. 2008-2009

[9] Al-Hayek SK. Contemporary life skills in line with integrated educational developments incorporated in the curricula of official Jordanian Universities. Mu'ta Magazine for Research, Human Sciences. 2010;**25**(4)

[10] Ayyad FIS, Al-Din HBMS. The effectiveness of a proposed conception to include some life skills in the technology course for the tenth grade in Palestine. Al-Aqsa University Journal (Humanities Series). 2010;**14**(1)

[11] Slav B. Quality of life from the perspective of positive psychology (analytical study). Journal of Social Studies and Research - Valley University. 2014;(08):402-432

[12] Al-Ajmi SBR. Quality of life and its relationship to the future orientation among students of the Faculty of Graduate Studies at Naif Arab University for Security Sciences « Factor Study » [PhD thesis] specialization in criminal psychology – Naif Arab University for Security Sciences, College of Social and Administrative Sciences, Department of Psychology, published. 2015

[13] El-Sherbiny KM, Walid AK. Recent Trends in Special Education Research and Studies. 1st ed. Alexandria: Publisher Dar Al-Wafa for the World of Printing and Publishing; 2013

[14] Hala YA-B. Life skills for people with mental inertia from students of the

faculty of Arts at Al-Qadisiyah University Uruk for Humanities. 2015;**07**(01):375-420

[15] Somaya TJ, Dalia KAW. Quality of life in light of some multiple intelligences among secondary school students from different disciplines, quality of life in light of some multiple intelligences among secondary school students from different disciplines. Arab Studies in Education and Psychology (ASEP). 2012;**Part I**(22):68-105. Third Research

[16] Muhammad AKA-M. Quality of life as a predictor of future anxiety among students of the Faculty of Education and Arts at the Northern Border University. Taibah University Journal of Educational Sciences. 2013;**10**(1):33-49

[17] Mohammadi F. Obstacles to the quality of family life, the Second National Forum on Communication and Quality of Life in the Family on April 9-10. 2013

[18] Amal IAB. 2015. Psychological Stress Experienced by Diabetic Patients and its Relation to Quality of Life. A supplementary study for a Master's degree in Education (Educational Guidance), Sudan University of Science and Technology. The study is titled "Psychological Stresses and its Relation to Quality of Life among Diabetics in Omdurman Locality."

[19] Al-Omari JF. The extent of public Jordanian university students' awareness of life skills in light of the knowledge economy, psychological and educational studies, laboratory for developing psychological and educational practices, issue 10/June. 2013. pp. 103-128.

[20] Ahmed Hussein AM, Doaa MA-M. Life Skills. 1st ed. Cairo: Dar Al-Sahab for Publishing and Distribution; 2007

[21] Judah A, Jarad HA. Predicting happiness in light of hope and optimism among a sample of Al-Quds Open University students. Journal of Al-Quds Open University for Research and Studies. 2011;**24**(2):129-162

[22] Al-Razi MABAQ. Mukhtar Al-Sahah. Lebanon: Department of Dictionaries in Lebanon, Library of Lebanon; 1989

[23] Ali MK, Abdul Khaleq NA-B. Quality of life among Omani and Libyan university students "A comparative cultural study". Journal of Humanities and Social Sciences. 2004;(20):67-87, September 2015 (The Scientific Journal of Arab Open Academy in Denmark, semi-annual peer-reviewed session)

[24] Huda HOA-Z. Happiness from an Islamic perspective, a thesis for a doctorate in Islamic studies. Khartoum; 2004

[25] Naglaa MS. Communication skills in social service, theoretical and scientific foundations. 1st ed. Amman, Jordan: Dar Al-Thaqafa for Publishing and Distribution; 2012

[26] Doaa, the Sawi Mr. Hussein Adel. The perceived quality of life among a sample of university students and the effectiveness of the program Rushdi and Joudi in its development, a memorandum to obtain a master's degree in education, specializing in mental health. 2009

[27] Iman BSJ. The role of the life skills program in achieving family security. A Master's Thesis - Naif Arab University for Security Sciences, College of Social and Administrative Sciences. Department of Sociology, Specialization in Social Rehabilitation and Care; 2015

[28] Hassouna A. Developmental Psychology. 1st ed. International Publishing House, Al-Ahram; 2004

[29] A practical guide to designing quality programs. Enhancing Life Skills Among Young People. International Youth Foundation (IYF); 2014

[30] Inas AAA-H. The Role of Information Technology in Time Management for Managers of UNRWA Schools in Gaza Governorates and Ways to Activate it. Doctoral thesis in Management. Al-Jazeera University; 2011

[31] Basam ATI. Problem-Based Learning and Critical Thinking. Jordan: Dar Al-Maseera for Publishing and Distribution; 2009

[32] Ahmed H A-M'a, Du'aa MM. Life Skills. 1st ed. Cairo: Dar Al-Sahab for Publishing and Distribution; 2008

[33] Fahim MM. Life skills in secondary school and the path to creating a modern personality, Arab Republic of Egypt, Training. 2005. pp. 120-155

[34] UNICEF. Available from: https://www.unicef.org/%E2%80%A6re/Whatwhy.Skills.LifeSkill

Chapter 5

Multivariate Analysis in the Characterization and Classification of Soils

Oswaldo Eduardo Ramos Ramos and Leonardo Guzmán Alegría

Abstract

Soil is a fundamental natural resource in the balance for the ecosystems as well as for agriculture, food, and housing. The soil is very susceptible to changes in its structure due to contamination or degradation of anthropogenic origin. Therefore, its evaluation, whether for environmental purposes or as an agricultural or housing resource, must be carried out in depth. This evaluation comprises the analysis of multiple physical, physicochemical and chemical-biological parameters. However, due to these multiple parameters, the use of multivariate statistical methods becomes necessary. In this chapter, the soil data analysis was performed by the method of Principal Components Analysis for a reduction of dimensions and, to carry out a better interpretation of results. This method was applied to carry out a characterization and classification of soil samples. The analysis was performed with data obtained from soil samples from the Bolivian Altiplano. The results show the potential of the principal component of the method in processing data.

Keywords: principal component analysis, multivariate analysis, reduction of dimensions, Bolivian Altiplano, contamination

1. Introduction

Soil and water are fundamental natural resources in the balance of ecosystems as well as in agriculture, food, and housing being thus fundamental for life.

Soils are composed of two environments, biotic and abiotic. The first constituted of microorganisms while the second, abiotic, is composed of solid, liquid and gaseous phases. The two environments characterize a particular soil giving it its uniqueness.

Soil and natural waters are highly susceptible to big changes in their structure and composition due to anthropogenic degradation or contamination. Therefore, its evaluation, whether for environmental purposes or as agricultural or housing resources, must be carried out deeply. This evaluation comprises the analysis of multiple physical, physicochemical and chemical-biological parameters. For example, to know the fertility of the soils, it is important to analyze parameters such as: pH, Electrical Conductivity, Organic Matter, exchangeable cations and others. The evaluations are carried out by comparison of these parameters with values established in agricultural or environmental regulations. However, due to complexity and variety and a big number of parameters, its analysis may become tough. Because of this, the use of multivariate statistical methods becomes imperative. Thus, the multivariate analysis applied to the characterization and classification of soils and natural waters according to the field of study that is intended to be carried out, grows up in importance.

The Principal Components Analysis offers us an alternative for the characterization and classification of soils. The different soil samples or soil sampling points constitute the elements of a system, and the physicochemical parameters measured in these samples, the variables. Thus, we have a system with multiple elements and variables since generally, the sampling points and the variables are numerous. The data analysis can be quite complex because the representative points among the sampling points should be represented in multidimensional spaces. Even though the variables could be represented in one-dimensional (considering each variable) or two-dimensional (every two variables) spaces, this is neither practical nor objectively informative.

The analysis through Principal Components with reduction of dimensions offers us precisely an alternative. Since the set of multivariables can be reduced to a few compound variables, the analysis becomes feasible and, the conclusions are more objective.

However, not all data sets are susceptible to Principal Component Analysis (PCA). They must meet certain requirements, for example they should be comprised of numerical variables and, the correlations between the variables must be above an acceptable level. If these requirements are not met, although a PCA could be made, their results would not be valid. The compliance for these requirements is given by the correlation matrix, where correlations must be observed. The Kaiser-Meyer-Olkin Measure (KMO) of sampling adequacy is a statistic that indicates the amount of variance in the variables that can be explained with the PCA. It is somewhat similar to the coefficient of determination R^2 from a linear regression analysis. Kaiser proposed the following criteria for KMO [1]:

$0.9 \leq KMO \leq 1.0$ = Excellent sample adequacy.
$0.8 \leq KMO \leq 0.9$ = Good sample adequacy.
$0.7 \leq KMO \leq 0.8$ = Acceptable sample adequacy.
$0.6 \leq KMO \leq 0.7$ = Regular sample adequacy.
$0.5 \leq KMO \leq 0.6$ = Bad sample adequacy.
$0.0 \leq KMO \leq 0.5$ = Unacceptable sample adequacy.

Therefore, the KMO is required to be a value, at least, greater than 0.7 for the PCA to be acceptable. Bartlett's test of sphericity is a statistical test which null hypothesis is an identity matrix. The acceptance of the null hypothesis means that there are no correlations (Sig. > 0.05). On the other hand, the alternative hypothesis is a non-identity matrix. The acceptance of this hypothesis means that there are correlations (Sig. < 0.05) and thus, a PCA can be performed. The statistical evaluation was made by SPSS Software.

2. PCA in the analysis of soil samples

In a study carried out by Ramos Ramos et al. [2], water samples from the Bolivian altiplano were analyzed. The results of various elements and variables determined in the water are presented in **Table 1**.

Table 2 shows the correlation matrix of parameters. It can be seen that there are correlations between different variables, which is an indicative of underlying structures. In principle, a PCA would be feasible.

Location	Sample ID	Water type	EC (µS/cm)	pH	Eh (mV)	HCO_3^- (mg/L)	F (mg/L)	Cl^- (mg/L)	SO_4^{2-} (mg/L)	Ca^{2+} (mg/L)	Mg^{2+} (mg/L)	Na^+ (mg/L)	K^+ (mg/L)	Ionic balance (%)
Cayhuasi	CAP1	Mg–Na–Ca–HCO_3–SO_4	1,120	7.45	167	561.4	0.01	13.5	170.8	69.8	85 5	80.5	14.3	4.1
Soracachi 1	SOP1	Mg–Ca–Na–HCO_3–SO_4	912	7.33	165	427.1	0.24	19.7	145.4	55.1	62.3	56.8	6.1	–1.0
Soracachi 2	SOP2	Mg–Na–Ca–HCO_3–SO_4	936	7.15	184	417.4	0.01	21.0	158.8	56.3	64.8	67.6	5.0	1.4
Paria	PAP1	Na–Mg–HCO_3–Cl–SO_4	2,120	7.41	173	717.6	1.04	198.2	242.8	72.5	58.6	291.5	42.1	–0.7
Chusaqueri	CHP1	Ca–Na–Cl	1,480	7.52	160	153.8	0.03	294.9	85.6	156.3	20.4	88.3	8.2	1.9
Toledo 1	TOP1	Na–Ca–Cl–HCO_3	1,400	7.72	154	270.9	0.01	233.0	90.4	68.8	15.4	183.5	30.4	0.5
Toledo 2	TOP2	Na–Ca–Cl	3,860	7.38	170	263.6	0.01	926.7	183.3	206.9	55.4	421.0	28.0	–0.9
Kulliri	KUP1	Na–HCO_3–B–SO_4	850	7.75	160	297.8	0.11	61.9	97.0	27.2	3.4	135.0	19.1	–4.3
Copacabanita	COP1	Ca–Na–SO_4	1,390	7.08	190	190.4	0.47	25.5	562.5	234.5	16.4	125.5	6.9	8.6
Tolaloma	TOLP1	Ca–HCO_3–SO_4	660	7.61	158	266.0	0.01	26.8	105.6	98.7	11.1	33.0	6.9	0.7
Andamarca	ANP1	Na–Ca–HCO_3–SO_4	1,190	7.69	169	483.3	0.01	74.1	142.8	114.7	13.1	179.0	18.5	7.3
Avaroa	AVP1	Ca–Na–HCO_3	590	7.29	169	290.4	0.19	14.1	48.0	72.3	7.4	49.2	5.2	2.3
Pampa Aullagas	PAMP1	Na–Ca–Cl–HCO_3–SO_4	1,360	6.41	218	219.7	0.35	164.5	144.2	69.9	22.4	137.0	26.7	1.0
Quillacas 1	QUP1	Na–B–Cl–HCO_3	1,200	7.90	141	234.3	0.26	189.7	79.1	36.2	6.7	175.0	16.6	–2.3
Quillacas 2	QUP2	Na–Cl	810	6.55	210	61.0	0.20	93.7	66.8	11.5	4.9	123.8	21.4	–2.0
Quillacas 3	QUP3	Na–Cl–B–SO_4	410	6.74	205	36.6	0.13	35.9	36.8	10.0	3.6	44.6	17.5	7.0
Condo K2	CONP2	Na–Ca–HCO_3	660	7.55	162	234.3	0.32	42.0	55.6	52.0	8.2	109.4	9.9	13.0
Condo K4	CONP4	Na–Ca–HCO_3–B	570	7.55	165	227.0	0.29	30.4	48.7	43.1	7.4	51.0	8.9	–3.9

Location	Sample ID	Water type	EC (μS/cm)	pH	Eh (mV)	HCO_3^- (mg/L)	F (mg/L)	Cl^- (mg/L)	SO_4^{2-} (mg/L)	Ca^{2+} (mg/L)	Mg^{2+} (mg/L)	Na^+ (mg/L)	K^+ (mg/L)	Ionic balance (%)
Caraynacha	CARP1	Ca–Na–HCO_3	400	7.48	168	172.1	0.13	20.9	30.5	42.9	6.2	44.4	13.9	9.6
Llapallapani	LLAP1	Na–Ca–SO_4–HCO_3	210	6.97	195	43.9	0.05	7.1	40.1	11.5	4.3	14.6	6.2	–3.0
Challapata	CHAP1	Ca–Na–Mg–HCO_3–CI	650	6.84	197	135.5	0.15	47.0	35.7	55.3	15.3	44.3	6.5	10.0
Huancane	HUAP1	Na–Ca–HCO_3–Cl	870	7.13	185	292.9	0.14	78.8	46.7	70.1	13.8	130.5	7.2	13.0
Irukasa	IRP1	Na–Cl	4,630	7.35	173	593.1	0.01	1,219.7	9.1	39.6	7.2	1,333	26.9	16.1
Realenga	REP1	Na–Ca–Mg–SO_4–HCO_3	770	6.81	206	178.2	0.30	37.1	176.6	56.0	23.1	101.0	5.8	8.9
Pazña	PAZP2	Na–Ca–SO_4–HCO_3	1,270	7.19	180	300.2	0.67	72.1	282.6	114.3	20.5	211.5	23.6	14.3
Totoral	TOTP1	Na–Cl–NO_3–SO_4	1,450	6.87	198	98.8	0.55	168.8	131.1	51.3	10.4	208.0	22.8	0.5
Cayumalliri	CAYP1	Ca–Na–HCO_3–SO_4	640	6.70	212	205.0	0.07	33.9	72.1	64.5	15.1	35.0	3.0	–1.1
Sora Sora	SORP1	Al–Ca–Mg–SO_4	4,500	3.79	378	0.0	4.06	28.1	1,020.2	130.7	63.6	45.5	6.7	1.6
Chapana[r]	CHAO1	Ca–Mg–Na–HCO_3–SO_4	280	7.60	157	109.8	0.01	9.5	44.2	20.8	8.8	16.4	3.9	–7.7
Totoral[r]	TOR1	Ca–SO_4	2,580	3.88	369	0.0	0.51	104.0	1,546.9	333.5	40.4	28.2	9.2	–2.1
Avicaya[r]	AVR1	Ca–SO_4	2,530	3.10	417	0.0	1.09	106.9	1,261.1	255.3	36.8	56.7	6.6	–2.1
Pazna[r]	PAZR1	Ca–Na–SO_4	1,960	4.71	221	14.6	0.06	151.7	874.9	191.3	36.2	128.5	12.0	3.3

Below detection limit.
Samples ID end P pertain to wells.
Samples ID end R pertain to rivers.
Reproduced with permission from Springer Nature; Excerpt from [2].

Table 1.
Major ion composition in ground and surface water samples in the Poopó basin from the Bolivian Altiplano.

Correlaciones

		EC	pH	Eh	HCO_3	F	Cl	SO_4	Ca	Mg	Na	K
EC	Pearson correlation	1	−.287	.429*	.091	.531**	.709**	.505**	.493**	.419*	.625**	.342
	sig. (2-tailed)		.112	.014	.620	.002	.000	.003	.004	.017	.000	.055
	N	32	32	32	32	32	32	32	32	32	32	32
pH	Pearson correlation	−.287	1	−.569**	.378*	−.393*	.120	−.624**	−.361*	−.109	.133	.102
	sig. (2-tailed)	.112		.001	.033	.026	.514	.000	.043	.554	.468	.580
	N	32	32	32	32	32	32	32	32	32	32	32
Eh	Pearson correlation	.429*	−.569**	1	−.462**	.582**	−.107	.815**	.569**	.276	−.145	−.325
	sig. (2-tailed)	.014	.001		.008	.000	.559	.000	.001	.126	.428	.070
	N	32	32	32	32	32	32	32	32	32	32	32
HCO_3	Pearson correlation	.091	.378*	−.462**	1	−.228	.307	−.415*	−.262	.313	.475**	.456**
	sig. (2-tailed)	.620	.033	.008		.210	.087	.018	.148	.081	.006	.009
	N	32	32	32	32	32	32	32	32	32	32	32
F	Pearson correlation	.531**	−.393*	.582**	−.228	1	−.128	.523**	.225	.358*	−.098	−.033
	sig. (2-tailed)	.002	.026	.000	.210		.486	.002	.215	.044	.594	.860
	N	32	32	32	32	32	32	32	32	32	32	32
Cl	Pearson correlation	.709**	.120	−.107	.307	−.128	1	−.085	.123	.033	.897**	.497**
	sig. (2-tailed)	.000	.514	.559	.087	.486		.643	.501	.857	.000	.004
	N	32	32	32	32	32	32	32	32	32	32	32
SO_4	Pearson correlation	.505**	−.624**	.815**	−.415*	.523**	−.085	1	.832**	.400*	−.151	−.171
	sig. (2-tailed)	.003	.000	.000	.018	.002	.643		.000	.023	.408	.350
	N	32	32	32	32	32	32	32	32	32	32	32

Correlaciones												
		EC	pH	Eh	HCO_3	F	Cl	SO_4	Ca	Mg	Na	K
Ca	Pearson correlation	.493**	−.361*	.569**	−.262	.225	.123	.832**	1	.353*	−.057	−.107
	sig. (2-tailed)	.004	.043	.001	.148	.215	.501	.000		.048	.756	.559
	N	32	32	32	32	32	32	32	32	32	32	32
Mg	Pearson correlation	.419*	−.109	.276	.313	.358*	.033	.400*	.353*	1	−.060	.063
	sig. (2-tailed)	.017	.554	.126	.081	.044	.857	.023	.048		.746	.733
	N	32	32	32	32	32	32	32	32	32	32	32
Na	Pearson correlation	.625**	.133	−.145	.475**	−.098	.897**	.151	−.057	−.060	1	.512**
	sig. (2-tailed)	.000	.468	.428	.006	.594	.000	.408	.756	.746		.003
	N	32	32	32	32	32	32	32	32	32	32	32
K	Pearson correlation	.342	.102	−.325	.456**	−.033	.497**	−.171	−.107	.063	.512**	1
	sig. (2-tailed)	.055	.580	.070	.009	.860	.004	.350	.559	.733	.003	
	N	32	32	32	32	32	32	32	32	32	32	32

. Correlation is significant at the 0.05 level (2-tailed).. Correlation is significant at the 0.01 level (2-tailed).*

Table 2.
Correlations Matrix of water samples in Poopó Basin, Bolivian Altiplano.

The results of the KMO and Bartlett's sphericity are (**Table 3**):

KMO and Bartlett's Test		
Kaiser Meyer Olkin Measure of Sampling Adequacy		.478
Bartlett's Test of Sphericity	Approx. Chi-Square	349.409
	df	55
	Sig.	.000

Table 3.
KMO and Bartlett's sphericity results for water samples.

The KMO statistics has a value of 0.478 which indicates that the data is not suitable for PCA, although Bartlett's test of sphericity has a Sig of 0.000. This means that the data is not suitable for a PCA.

Table 4 correspond to an analysis of metals in soils samples carried out in the Bolivian altiplano, Poopó Basin, Bolivia [2]. There are 36 sampling points in which 16 parameters have been determined. Therefore, we have a system with 33 elements and 16 variables.

Table 5 shows the correlation matrix, where it can be seen that there are correlations between the different variables. KMO's and Bartlett's tests give the following results: The KMO statistics is 0.704 and indicates that the data is acceptable (**Table 6**) for performing a PCA. The Sig. of the Bartlett test is 0.000 and indicates that the alternative hypothesis is valid, thus, the correlation matrix is not an identity. Then, these values indicate that the data can be subjected to a PCA. Therefore, we proceed with the PCA.

Four main components have been extracted. The principal extracted components are presented in **Table 7**. The table shows that with four components, 85.13% of the variability would be explained.

Table 8 shows the rotated component matrix with the Varimax rotation method. According to this matrix, the representativeness of the main components with respect to the variables is determined. The highest correlation that the variable has with the main component was taken as a criterion.

$$\begin{aligned} PC1 &= Ni + Fe + Cu + Mn + Cr \\ PC2 &= B + Al + Si + Sr + Mo \\ PC3 &= Cd + Pb + Zn + P + As \\ PC4 &= Ni + Fe + Cu + Mn + Cr \end{aligned} \tag{1}$$

Table 9 shows the Component Score Coefficients Matrix, which are the coefficients of the variables in each PCA. For example, for Principal Component 1 (PC1), the following Eq. (2) applies:

$$\begin{aligned} PC1 = \; & -0.007 * Al + 0.078 * As - 0.184 * B - 0.125 * Cd + 0.145 * Cr + 0.206 * Cu \\ & + 0.259 * Fe + 0.220 * Mn - 0.106 * Mo + 0.278 * Ni + 0.009 * P - 0.041 * Pb \\ & - 0.051 * S - 0.031 * Si + 0.002 * Sr - 0.034 * Zn \end{aligned} \tag{2}$$

Similarly, the equations for the components PC2, PC3 and PC4 can be expressed.

No	Code	Al	As	B	Cd	Cr	Cu	Fe	Mn	Mo	Ni	P	Pb	S	Si	Sr	Zn
		mg/kg	mg/kg	mg/kg	mg/kg	mg/kg	mg/kg	mg/kg	mg/kg	mg/kg	mg/kg	mg/kg	mg/kg	mg/kg	mg/kg	mg/kg	mg/kg
1	COR 1H	13181.1	13.8	11.3	0.5	12.9	20.6	16708.9	233.7	0.2	8.0	532.5	35.6	556.9	429.9	42.5	86.8
2	COR 1C	11823.0	17.8	4.5	0.5	15.2	21.1	23063.1	474.0	0.3	10.1	337.6	40.9	300.5	519.2	32.5	84.2
3	COR 1P	12705.8	15.0	6.3	0.3	14.8	18.1	20914.0	370.3	0.3	9.5	345.5	35.1	161.0	459.4	26.3	85.2
4	COR 1AA	30996.3	38.5	15.0	0.9	40.2	48.6	57156.4	1029.9	0.8	27.4	976.4	88.3	581.1	1183.3	55.5	232.5
5	COR 2C	7945.6	28.0	8.9	0.7	10.2	20.1	15650.0	262.1	0.3	4.4	298.4	61.3	241.5	509.1	30.1	77.8
6	COR 2P	11231.6	31.2	6.9	0.7	10.9	21.3	16403.3	299.2	0.5	7.5	285.8	104.4	424.9	98.4	31.4	76.3
7	COR 2AA	8736.8	26.1	12.1	0.5	10.0	20.9	15818.3	332.4	0.3	5.8	407.7	48.7	373.7	449.3	37.8	99.5
8	COR 3H	10296.9	27.8	13.1	0.6	10.5	17.6	16405.1	394.2	0.5	6.8	457.7	40.3	348.8	535.2	33.3	81.4
9	COR 3C	4372.9	16.2	4.7	0.2	5.9	8.7	10264.0	192.5	0.3	3.4	218.9	20.4	168.4	375.9	15.6	54.5
10	COR 3P	6899.8	20.8	5.4	0.4	8.9	12.8	14938.4	281.6	0.3	5.1	279.2	30.6	153.6	455.2	21.8	67.0
11	COR 3AA	6910.1	21.5	6.7	0.5	9.1	16.0	15030.0	303.4	0.2	5.1	274.4	37.2	181.8	504.6	22.7	77.8
12	VM 1H	42276.0	30.1	40.0	1.2	27.1	27.8	22425.4	546.1	2.8	7.7	910.0	85.9	506.3	983.3	51.2	159.4
13	VM 1C	25081.6	23.0	27.5	0.3	17.4	11.5	16043.0	365.0	0.4	4.7	381.1	26.0	264.1	985.4	44.1	58.0
14	VM 2H	43636.7	17.3	29.5	0.3	29.2	18.8	22690.3	448.1	0.3	10.3	469.2	28.0	227.9	1138.1	39.1	78.0
15	VM 2P	40756.0	19.3	27.5	0.3	28.2	18.7	22097.1	428.5	0.4	8.8	449.3	25.6	220.4	1042.0	39.3	72.3
16	VM 3H	5509.4	9.3	8.3	0.4	5.2	9.4	8763.7	144.1	0.5	3.7	198.2	15.8	170.1	558.5	18.5	95.6
17	VM 3C	11299.4	13.3	3.8	0.6	15.3	22.9	23362.0	454.6	0.5	10.8	352.7	35.4	162.2	460.5	27.1	81.4
18	VM 3P	12348.8	15.9	4.8	0.4	16.3	25.5	25623.8	497.5	0.3	11.5	393.5	41.5	180.9	458.4	28.9	91.7
19	VM 4H	12362.6	19.7	20.5	0.8	11.1	21.8	17023.4	292.1	0.6	8.0	533.1	39.5	424.4	1395.9	51.3	203.0
20	VM 4C	25008.0	40.2	14.5	0.7	30.6	41.1	46040.2	928.5	0.9	18.9	813.9	70.8	398.4	1109.0	63.3	146.4
21	VM 4P	8614.9	18.5	23.3	0.8	9.1	16.6	18128.2	627.7	0.4	7.5	457.9	32.4	584.8	617.7	35.2	179.4
22	VM 5H	6887.9	28.9	11.8	0.9	10.0	22.1	238.7	503.4	0.3	7.8	362.5	70.2	844.3	487.6	25.5	130.2

No	Code	Al	As	B	Cd	Cr	Cu	Fe	Mn	Mo	Ni	P	Pb	S	Si	Sr	Zn
		mg/kg	mg/kg	mg/kg	mg/kg	mg/kg	mg/kg	mg/kg	mg/kg	mg/kg	mg/kg	mg/kg	mg/kg	mg/kg	mg/kg	mg/kg	mg/kg
23	VM 5C	7839.3	28.4	13.3	0.9	10.8	22.5	21060.7	505.6	0.3	7.3	371.0	73.0	818.4	488.2	26.0	137.5
24	VM 6C	7941.5	19.0	9.3	0.4	8.9	14.8	16193.1	510.6	0.3	6.5	328.9	43.9	360.1	575.9	24.6	88.5
25	VM 6P	6511.7	27.2	8.6	1.1	9.3	21.1	21475.7	544.7	0.2	6.7	364.7	53.6	414.0	506.9	19.5	161.2
26	POO 1C	10751.3	19.8	5.7	0.9	12.3	20.3	19361.9	424.3	0.5	8.6	348.3	43.4	191.3	633.8	27.3	90.8
27	POO 1P	13660.9	20.6	7.1	0.5	14.4	21.0	22414.2	485.6	0.6	9.4	348.3	33.7	181.2	606.7	33.8	76.1
28	POO 2C	10290.1	15.7	13.5	1.0	10.3	14.0	14995.4	340.5	0.4	5.2	337.6	27.9	179.9	516.9	36.1	117.9
29	POO 2P	10725.5	19.3	6.7	0.4	12.2	15.7	17443.0	386.6	0.2	6.8	295.3	30.9	214.1	552.8	29.0	68.3
30	POO 3C	7681.3	27.0	8.4	2.1	9.9	19.1	17784.7	297.0	0.3	6.4	688.2	138.7	300.1	496.8	30.3	246.6
31	POO 3P	9274.6	21.2	12.1	1.5	11.0	20.8	17636.1	337.4	0.4	6.0	803.9	95.2	289.3	517.9	36.9	193.8
32	POO 4C	7099.8	14.3	9.9	0.3	10.9	12.2	14469.9	286.8	0.7	7.5	250.7	24.8	171.4	440.3	26.0	51.7
33	POO 4C (Alta)	9838.7	17.5	11.7	0.4	10.3	13.0	14625.4	277.9	0.3	5.1	331.2	24.4	182.9	503.2	46.3	47.7
34	POO 4P	10407.3	16.3	11.7	0.6	11.3	15.1	16412.7	359.6	0.3	6.5	348.0	38.3	292.0	664.4	33.9	69.1
35	POO 4 AA	25849.4	14.6	28.9	0.3	18.6	12.7	14794.3	311.3	0.3	5.6	315.7	27.1	583.4	1439.7	54.3	57.0
36	PUÑ 4P	22049.2	16.5	21.2	0.2	15.2	9.3	12327.1	348.1	0.5	4.8	317.3	16.6	232.0	1308.1	43.7	45.5
37	Referencia	33277.5	18.4	23.7	3.9	74.5	356.4	26390.7	491.4	0.4	300.2	969.5	101.5	1841.5	501.3	81.0	744.7

Table 4.
Metals and trace elements analysis in soils from Poopó Basin, in Bolivian Altiplano [3].

Correlations

		Al	As	B	Cd	Cr	Cu	Fe	Mn	Mo	Ni	P	Pb	S	Si	Sr	Zn
Al	Pearson Correlation	1	,207	,796**	−,115	,876**	,353*	,452**	,400*	,512**	,412*	,505**	,012	,092	,716**	,651**	,039
	Sig. (2-tailed)		,227	,000	,504	,000	,035	,006	,016	,001	,012	,002	,946	,595	,000	,800	,819
	N	36	36	36	36	36	36	36	36	36	36	36	36	36	36	36	36
As	Pearson Correlation	,207	1	,124	,449**	,421*	,726**	,528**	,635**	,334*	,508**	,595**	,700**	,481**	,106	,354*	,503**
	Sig. (2-tailed)	,227		,472	,006	,011	,000	,001	,000	,047	,002	,000	,000	,003	,539	,034	,002
	N	36	36	36	36	36	36	36	36	36	36	36	36	36	36	36	36
B	Pearson Correlation	,796**	,124	1	,003	,528**	,056	,083	,184	,530**	,028	,414*	−,034	,327	,732**	,642**	,136
	Sig. (2-tailed)	,000	.472		,988	,001	,745	,632	,282	,001	,871	,012	,845	,052	,000	,000	,429
	N	36	36	36	36	36	36	36	36	36	36	36	36	36	36	36	36
Cd	Pearson Correlation	−,115	,449**	,003	1	−,030	,355*	,133	,189	,238	,099	,581**	,823**	,310	−,118	,057	,826**
	Sig. (2-tailed)	,504	,006	,988		,864	,033	,438	,271	,163	,567	,000	,000	,066	,491	,743	,000
	N	36	36	36	36	36	36	36	36	36	36	36	36	36	36	36	36
Cr	Pearson Correlation	,876**	,421*	,528**	−,030	1	,700**	,788**	,705**	,443**	,786**	,660**	,167	,154	,638**	,685**	,218
	Sig. (2-tailed)	,000	,011	,001	,864		,000	,000	,000	,007	,000	,000	,331	,368	,000	,000	,202
	N	36	36	36	36	36	36	36	36	36	36	36	36	36	36	36	36
Cu	Pearson Correlation	,353*	,726**	,056	,355*	,700**	1	,856**	,834**	,366*	,898**	,760**	,553**	,396*	,225	,492**	,587**
	Sig. (2-tailed)	,035	,000	,745	,033	,000		,000	,000	,028	,000	,000	,000	,017	,187	,002	,000
	N	36	36	36	36	36	36	36	36	36	36	36	36	36	36	36	36
Fe	Pearson Correlation	,452**	,528**	,083	,133	,788**	,856**	1	,825**	,272	,915**	,665**	,298	,091	,348*	,521**	,414*

Correlations

		Al	As	B	Cd	Cr	Cu	Fe	Mn	Mo	Ni	P	Pb	S	Si	Sr	Zn
	Sig. (2-tailed)	,006	,001	,632	,438	,000	,000		,000	,109	,000	,000	,077	,596	,037	,001	,012
	N	36	36	36	36	36	36	36	36	36	36	36	36	36	36	36	36
Mn	Pearson Correlation	,400*	,635**	,184	,189	,705**	,834**	,825**	1	,310	,857**	,622**	,309	,401*	,336*	,442**	,470**
	Sig. (2-tailed)	,016	,000	,282	,271	,000	,000	,000		,065	,000	,000	,067	,015	,045	,007	,004
	N	36	36	36	36	36	36	36	36	36	36	36	36	36	36	36	36
Mo	Pearson Correlation	,512**	,334*	,530**	,238	,443**	,366*	,272	,310	1	,224	,564**	,271	,152	,296	,413*	,257
	Sig. (2-tailed)	,001	,047	,001	,163	,007	,028	,109	,065		,190	,000	,110	,376	,079	,012	,130
	N	36	36	36	36	36	36	36	36	36	36	36	36	36	36	36	36
Ni	Pearson Correlation	,412*	,508**	,028	,099	,786**	,898**	,915**	,857**	,224	1	,629**	,299	,238	,321	,476**	,421*
	Sig. (2-tailed)	,012	,002	,871	,567	,000	,000	,000	,000	,190		,000	,077	,163	,056	,003	,010
	N	36	36	36	36	36	36	36	36	36	36	36	36	36	36	36	36
P	Pearson Correlation	,505**	,595**	,414*	,581**	,660**	,760**	,665**	,622**	,564**	,629**	1	,622**	,352*	,395*	,648**	,744**
	Sig. (2-tailed)	,002	,000	,012	,000	,000	,000	,000	,000	,000	,000		,000	,035	,017	,000	,000
	N	36	36	36	36	36	36	36	36	36	36	36	36	36	36	36	36
Pb	Pearson Correlation	,012	,700**	–,034	,823**	,167	,553**	,298	,309	,271	,299	,622**	1	,414*	–,147	,137	,701**
	Sig. (2-tailed)	,946	,000	,845	,000	,331	,000	,077	,067	,110	,077	,000		,012	,392	,425	,000
	N	36	36	36	36	36	36	36	36	36	36	36	36	36	36	36	36
S	Pearson Correlation	,092	,481**	,327	,310	,154	,396*	,091	,401*	,152	,238	,352*	,414*	1	,168	,284	,453**
	Sig. (2-tailed)	,595	,003	,052	,066	,368	,017	,596	,015	,376	,163	,035	,012		,328	,093	,006

Correlations

		Al	As	B	Cd	Cr	Cu	Fe	Mn	Mo	Ni	P	Pb	S	Si	Sr	Zn
	N	36	36	36	36	36	36	36	36	36	36	36	36	36	36	36	36
Si	Pearson Correlation	,716**	,106	,732**	−,118	,638**	,225	,348*	,336*	,296	,321	,395*	−.147	,168	1	,737**	,163
	Sig. (2-tailed)	,000	,539	,000	,491	,000	,187	,037	,045	,079	,056	,017	,392	,328		,000	,342
	N	36	36	36	36	36	36	36	36	36	36	36	36	36	36	36	36
Sr	Pearson Correlation	,651**	,354*	,642**	,057	,685**	,492**	,521**	,442**	,413*	,476**	,648**	,137	,284	,737**	1	,253
	Sig. (2-tailed)	,000	,034	,000	,743	,000	,002	,001	,007	,012	,003	,000	,425	,093	,000		,136
	N	36	36	36	36	36	36	36	36	36	36	36	36	36	36	36	36
Zn	Pearson Correlation	,039	,503**	,136	,826**	,218	,587**	,414*	,470**	,257	,421*	,744**	,701**	,453**	,163	,253	1
	Sig. (2-tailed)	,819	,002	,429	,000	,202	,000	,012	,004	,130	,010	,000	,000	,006	,342	,136	
	N	36	36	36	36	36	36	36	36	36	36	36	36	36	36	36	36

***Correlation is significant at the 0.01 level (2-tailed).*Correlation is significant at the 0.05 level (2-tailed).*

Table 5.
Sample correlation matrix of metal and trace elements in soil samples.

KMO and Bartlett's Test		
Kaiser Meyer Olkin Measure of Sampling Adequacy		.704
Bartlett's Test of Sphericity	Approx. Chi-Square	725.706
	df	120
	Sig.	.000

Table 6.
KMO and Bartlett's sphericity results for soil samples.

Total variance explained									
Component	Initial eigenvalues			Extraction sums of squared loadings			Rotation sums of squared loadings		
	Total	% of variance	Cumulative %	Total	% of variance	Cumulative %	Total	% of variance	Cumulative %
1	7.584	47.401	47.401	7.584	47.401	47.401	4 732	29.578	29.578
2	3.192	19.948	67.350	3.192	19.948	67.350	4.070	25.435	55.013
3	1.905	11.904	79.254	1.905	11.904	79.254	3.639	22.743	77.756
4	.941	5.880	85.134	.941	5.880	85.134	1.181	7.379	85.134
5	.735	4.596	89.731						
6	.466	2.913	92.643						
7	.381	2.381	95.024						
8	.289	1.805	96.829						
9	.164	1.025	97.854						
10	.110	.690	98.544						
11	.074	.462	99.006						
12	.065	.405	99.411						
13	.049	.304	99.715						
14	.029	.181	99.896						
15	.011	.072	99.968						
16	.005	.032	100.000						
Extraction Method: Principal Component Analysis									

Table 7.
Extraction of principal components (PC) with a total explained variance of 85.13%.

According to this matrix of score coefficients and their corresponding equations, the score values of the main components are obtained for the 36 sampling points. These are shown in **Table 10**.

After these scores, which represent the new composite reduced variables, different graphical representations and interpretations can be made. For example, the plotting of each principal component with respect to the sampling points (**Figure 1**).

Figure 1a shows that in the sampling points there is a certain homogeneous distribution of PC1 with respect to the points, except for points 4 (Code sample, CORR

Rotated component matrix[a]				
	Component			
	1	2	3	4
Al	.320	.874	−.056	−.101
As	.523	.083	.547	.314
B	−.116	.945	.057	.197
Cd	−.011	−.068	.946	.056
Cr	.719	.647	.016	−.065
Cu	.861	.132	.412	.125
Fe	.928	.203	.155	−.104
Mn	.850	.199	.191	.233
Mo	.099	.609	.432	−.250
Ni	.956	.142	.115	.046
P	.500	.470	.658	.001
Pb	.220	−.078	.885	.129
S	.119	.161	.328	.874
Si	.210	.825	−.145	.166
Sr	.376	.746	.093	.163
Zn	.280	.073	.809	.214

Extraction Method: Principal Component Analysis.
Rotation Method: Varimax with Kaiser Normalization.
[a]Rotation converged in 5 iterations.

Table 8.
Rotated component matrix.

Matriz de coeficiente de puntuación de componente				
	Componente			
	1	2	3	4
Al	−,007	,236	−,027	−,140
As	,078	−,050	,079	,192
B	−,184	,308	,020	,145
Cd	−,125	−,003	,351	−,124
Cr	,145	,108	−,066	−,117
Cu	,206	−,076	,018	,017
Fe	,259	−,058	−,043	−,168
Mn	,220	−,063	−,084	,166
Mo	−,106	,200	,213	−,365
Ni	,278	−,089	−,093	−,008
P	,009	,093	,193	−,161

Matriz de coeficiente de puntuación de componente				
	Componente			
	1	2	3	4
Pb	−,041	−,043	,282	−,044
S	−,051	,000	−,057	,803
Si	−,031	,222	−,099	,150
Sr	,002	,178	−,032	,101
Zn	−,034	−,009	,231	,048

Método de extracción: análisis de componentes principales.
Método de rotación: Varimax con normalización Kaiser.
Puntuaciones de componente.

Table 9.
Component score coefficients matrix.

Sample	CP_1	CP_2	CP_3	CP_4
1	−0.21118	−0.0777	−0.33467	0.80961
2	0.6651	−0.67391	−0.53521	−0.20777
3	0.39744	−0.60917	−0.64339	−0.954
4	4.07162	0.56719	0.76641	0.55046
5	−0.34113	−0.59646	0.24423	−0.13876
6	−0.07872	−0.94299	0.7625	0.27764
7	−0.18808	−0.3646	0.02199	0.54976
8	−0.16801	−0.19284	0.04181	0.26754
9	−0.79917	−0.90177	−0.74427	−0.65343
10	−0.33788	−0.78329	−0.45122	−0.68191
11	−0.27713	−0.79168	−0.32166	−0.4073
12	−0.67221	3.28014	2.51067	−1.29964
13	−0.61199	1.3241	−0.81815	0.22432
14	0.33766	1.77831	−1.07201	−0.46932
15	0.23938	1.63463	−0.98365	−0.54454
16	−1.18136	−0.4418	−0.4608	−0.8548
17	0.68138	−0.72139	−0.37866	−1.24838
18	1.03018	−0.79119	−0.53675	−0.89173
19	−0.59117	1.09146	0.47815	0.82755
20	2.92182	0.70194	0.35112	0.21752
21	−0.32097	0.11616	0.21485	1.62766
22	−0.52482	−0.6866	0.55143	2.96642
23	0.00703	−0.71324	0.53004	2.46939
24	−0.07964	−0.61843	−0.43321	0.45935

Sample	CP_1	CP_2	CP_3	CP_4
25	0.13395	−1.0284	0.67803	0.67771
26	0.16106	−0.54494	0.12387	−0.82888
27	0.54161	−0.31881	−0.42751	−0.80632
28	−0.6818	−0.06185	0.23561	−0.69627
29	0.06159	−0.56759	−0.68723	−0.29851
30	−0.80934	−0.64774	3.33121	−0.66866
31	−0.59454	−0.14172	2.187	−0.70855
32	−0.38801	−0.34591	−0.64539	−0.98714
33	−0.47597	0.01196	−0.67739	−0.40583
34	−0.32046	−0.14616	−0.38936	−0.05153
35	−0.83523	1.86013	−1.28317	1.80646
36	−0.76102	1.34416	−1.20524	0.07187

Table 10.
Sampling point scores for principal components.

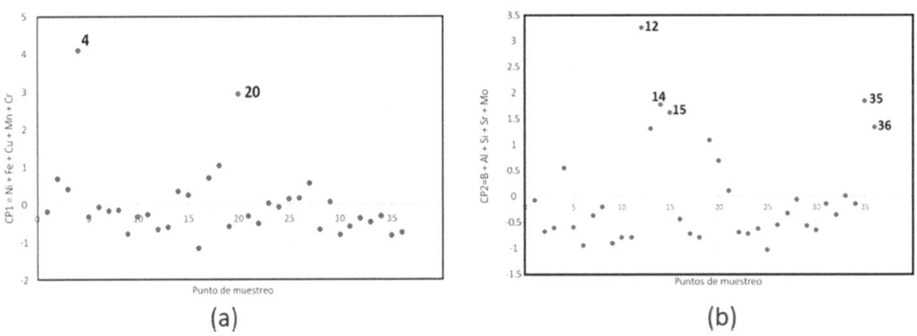

Figure 1.
Representation of the main components PC1 and PC2 with respect to the sampling points.

1AA) and 20 (Code sample, VM 4C). This indicates that there are high outliers in PC1 (Ni, Fe, Cu, Mn and Cr). In environmental terms, special attention is required at these sampling points. The points in question are (**Table 11**):

No	Code Soil	Cr mg/kg	Cu mg/kg	Fe mg/kg	Mn mg/Kg	Ni mg/kg
4	COR 1AA	40.2	48.6	5756.4	1029.9	27.4
20	VM 4C	30.6	41.1	46040.2	928.5	18.9
	Max	40.2	48.6	57156.4	1029.9	27.4
	Min	5.2	8.7	238.7	144.1	3.4
	Mean	14.3	19.3	19049.5	411.18	7.9

Table 11.
Inhomogeneous distribution in soil samples.

The samples have high values in these metals. Particularly, sample number 4 presents the maximum values and, sample number 20 is well above the average values. Similar analysis can be done for the other principal components.

Regarding structure, the samples numbers 4 and 20 form a group with relatively high values of PC1. Another group is formed by the rest of the points. **Figure 1b** represents the distribution of PC2 (B, Al, Si, Sr and Mo). It can also be seen that there are atypical points forming a group with relatively high values in this component. These are sampling points 12, 14, 15, 35 and 36. The rest of the points form a group with a homogeneous distribution in relation to PC2.

Figure 2a shows that the behavior of the PC3 is also homogeneous, with the exception of sampling points 12, 30, and 31 that would have high outliers in Cd, Pb, Zn, P and As. Regarding structure, sampling points 12, 30 and 31 form a group with high values of PC3, and the other group formed by the rest of the sampling points. **Figure 2b** represents the distribution of sulfur in the sampling points. It can be seen that one group would be made up of sampling points 21, 22, 23 and 35 with relatively high sulfur values, and the other group made up of the rest of the points.

Figure 3 is the representation of PC1 against PC2. There are three groups observed; the first group formed by sampling points 12, 13, 14, 15, 19, 35 and 36, with high values of PC2 (B, Al, Si, Sr and Mo) with respect to their values of PC1 (Ni, Fe, Cu,

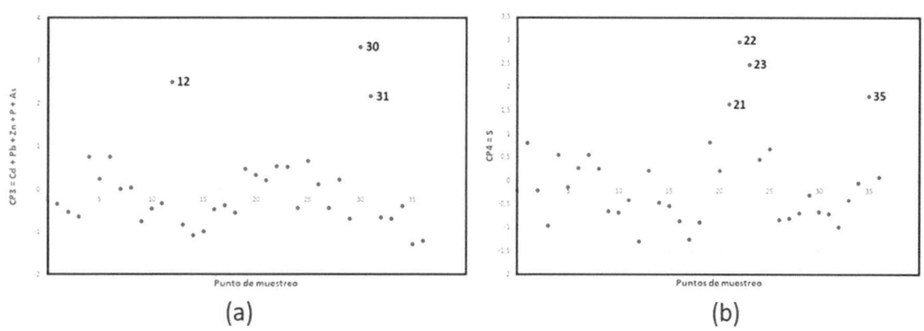

Figure 2.
Representation of the main components PC3 and PC4 with respect to the sampling points.

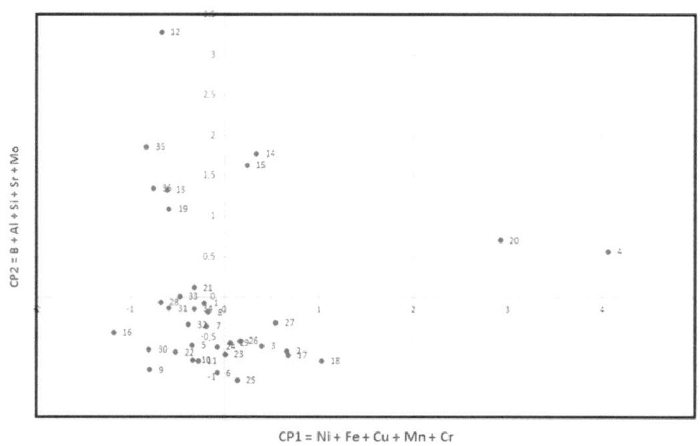

Figure 3.
Representation of the main components PC2 against PC1.

Mn and Cr); a second group made up of sampling points 4 and 20 which would have high values of PC1 in relation to PC2; and a third group made up of the rest of the sampling points which would have an homogeneous distribution in PC2 and PC1. Again, from the environmental point of view, the sampling points of the first and second groups should be analyzed more carefully.

Figure 4 shows the distribution of the samples in relation to the main components PC1 against PC3. While the PC3 in most of the sampling points does not show variability except for points 12, 30 and 31; PC1 is highly distributed with the greatest variability in the sampling points.

In the same way, graphical representations of the rest of the combinations of the main components can be made (**Figures 5** and **6**).

In the following example, two types of soils have been considered. The characteristics of both are different, and this fact will allow us to see the ability of the main components to characterize and classify soils [4].

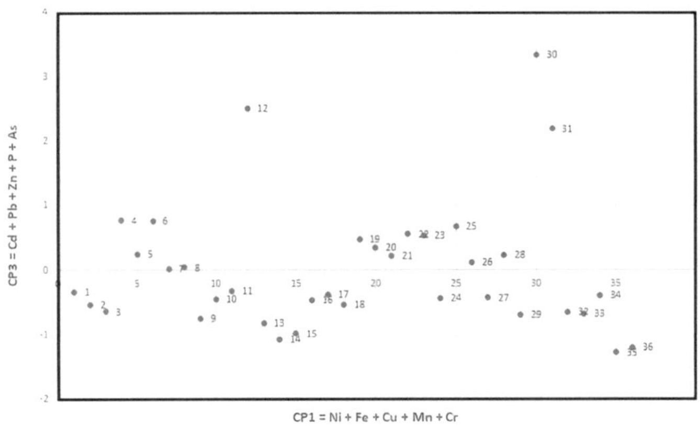

Figure 4.
Representation of the main components PC3 against PC1.

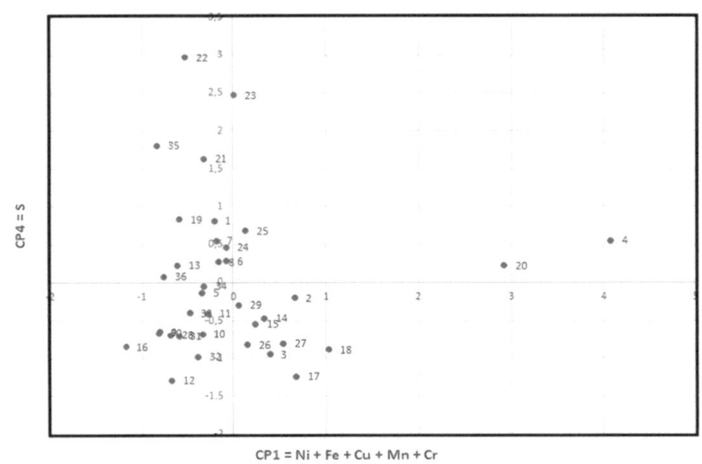

Figure 5.
Representation of the main components PC4 against PC1.

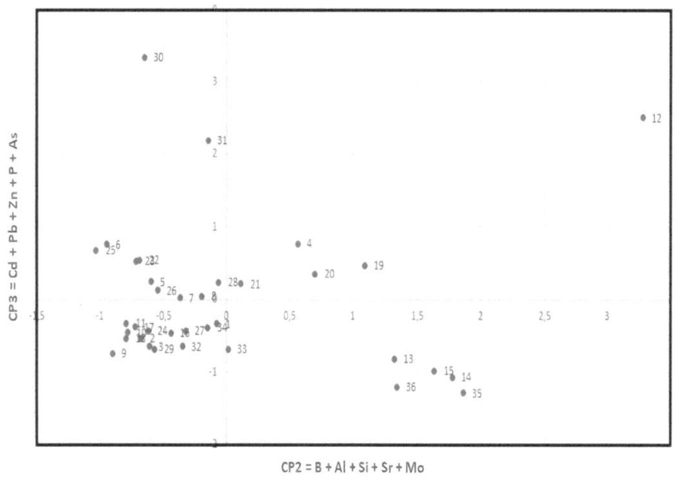

Figure 6.
Representation of the main components PC3 against PC2.

The following chemical parameters have been considered: pH in H_2O, pH in KCl solution, Electrical Conductivity (EC), Change acidity, Total Nitrogen, Organic Matter, Assimilable Phosphorus and Exchangeable Cations (Ca^{2+}, Mg^{2+}, Na^{+}, K^{+}).

The first group of samples comes from the inter-Andean valley of the Municipality of Inquisivi – Yamora, which is located between the coordinates: 66°43′29″ and 67°17′58″ West longitude; 15°47′34″ and 17°18′20″ South latitude and at an average altitude of 2840 m (a.s.l.). The second sample comes from the Northern Altiplano Viacha Municipality, located between the coordinates: 68°16′56″ and 68°22′72″ West longitude and 16°32′39″ and 16°54′44″ latitude, with an average altitude of 4070 m (a.s.l.), both in La Paz, Bolivia [4].

The ten soil samples have been taken in the Yamora community, and another 10 soil samples from the Viacha community. The mentioned 11 parameters have been analyzed. The evaluation does not take into account the environmental conditions of Yamora or Viacha. It is only carried out based on the chemical parameters for the evaluation of fertility from the chemical point of view of the soils (**Table 12**).

Location	pH (H_2O)	pH (KCl)	CE	H-Al	% MO	% N	Na	K	Ca	Mg	P
Yamora	6.75	5.8	0.075	0.0329	3.4	0.28	0.128	0.688	17.761	2.548	273.916
Yamora	6.76	4.98	0.075	0.0609	3.2	0.30	0.128	0.688	17.760	2.577	255.876
Yamora	6.72	5.73	0.068	0.0339	3.3	0.31	0.134	0.688	18.331	2.636	250.994
Yamora	6.76	5.89	0.074	0.0082	3.4	0.32	0.134	0.655	18.674	2.684	257.253
Yamora	6.73	5.89	0.072	0.0329	3.2	0.30	0.134	0.655	17.455	2.518	246.810
Yamora	6.79	5.92	0.068	0.0329	3.4	0.32	0.146	0.655	17.799	2.548	266.998
Yamora	6.79	5.37	0.069	0.0391	3.4	0.3	0.128	0.688	17.874	2.587	253.086
Yamora	6.8	5.84	0.073	0.0329	3.4	0.3	0.134	0.655	17.074	2.450	259.345
Yamora	6.83	6.01	0.072	0.0349	3.4	0.3	0.134	0.655	18.027	2.606	261.420

Location	pH (H_2O)	pH (KCl)	CE	H-Al	% MO	% N	Na	K	Ca	Mg	P
Yamora	6.82	5.95	0.072	0.0329	3.1	0.3	0.134	0.622	17.341	2.479	275.347
Viacha	8.54	7.13	0.727	0.0934	0.7	0.086	4.663	0.459	5.385	4.008	16.010
Viacha	8.72	7.16	0.732	0.0934	0.7	0.096	5.012	0.491	5 155	4.066	15.870
Viacha	8.78	7.12	0.736	0.1054	0.7	0.101	4.605	0.426	5.042	3.988	18.241
Viacha	8.74	7.11	0.735	0.1054	0.5	0.093	4.605	0.426	5.080	3.988	19.287
Viacha	8.81	7.16	0.737	0.0934	0.6	0.092	4.663	0.426	5.118	4.027	19.845
Viacha	8.78	7.12	0.738	0.1054	0.7	0.091	4.663	0.459	5.118	4.027	20.612
Viacha	8.94	7.01	0.731	0.1091	0.7	0.093	4.605	0.426	5.080	3.959	17.683
Viacha	8.69	6.78	0.733	0.0813	0.6	0.093	4.663	0.426	5.309	3.802	18.241
Viacha	8.49	7.14	0.732	0.0818	0.7	0.093	5.129	0.491	5.233	3.978	18.311
Viacha	8.81	7.14	0.780	0.0813	0.5	0.093	5.246	0.491	4.889	3.939	16.010

Reproduced with permission from Revista Boliviana de Química; Excerpt from [4].

Table 12.
Results of the analysis of parameters in samples from Yamora and Viacha.

PCA was also performed and the correlation matrix is shown in **Table 13**.

In the correlation matrix, high correlations between the variables are observed, the KMO with 0.865 and a Bartlett Significance of 0.000 indicate that the reduction of dimensions by principal components is feasible and adequate (**Table 14**). Therefore, we proceeded to obtain two main components (**Table 15**) and the rotated component matrices and component score coefficients for the samples (**Table 16**) with the application of Varimax rotation and Kaiser normalization.

The rotated component matrix shows that there is a structure. A group of parameters that have a positive correlation with the principal components, Group 1 (PC1): pH in KCl, Na^+, CE, Mg^{2+}, pH in H2O and H-Al with positive correlation. There is another group of parameters that have a negative correlation, Group 2 (PC2): N, MO, P, Ca^{2+} and K^+. This leads to a competition between these groups of parameters in the soil. If Group 1 overlaps Group 2, the soil would have high pH and EC values, high Na^+ and Mg^{2+} contents, and positive values of the main components, poor exchange content of OM, P and N. It means that there is an unfavorable soil for agriculture purposes. However, if Group 2 of parameters overlaps Group 1, then the soil is rich in OM, P and N. This means that the soil is more suitable for agriculture purposes, and it would have negative values of the main components.

The score coefficient matrix of the components generates the functions of PC1 and PC2:

$$PC1 = 0.011 * pH(H2O) + 1.195 * pH(KCl) + 0.106 * CE - 1.184 * H_{Al} - 0.041 * OM - 0.007 * N + 0.185 * Na - 0.174 * K - 0.047 * Ca + 0.095 * Mg - 0.045 * P \quad (3)$$

Correlaciones												
		pHenH2O	**pHenKCl**	**CE**	**H_Al**	**MO**	**N**	**Na**	**K**	**Ca**	**Mg**	**P**
pHenH2O	Correlación de Pearson	1	.944**	.996**	.945**	−.994**	−.992**	.991**	−.978**	−.996**	.991**	−.993**
	Sig. (bilateral)		.000	.000	.000	.000	.000	.000	.000	.000	.000	.000
	N	20	20	20	20	20	20	20	20	20	20	20
pHenKCl	Correlación de Pearson	.944**	1	.947**	.836**	−.941**	−.941**	.948**	−.940**	−.947**	.947**	−.941**
	Sig. (bilateral)	.000		.000	.000	.000	.000	.000	.000	.000	.000	.000
	N	20	20	20	20	20	20	20	20	20	20	20
CE	Correlación de Pearson	.996**	.947**	1	.936**	−.998**	−.997**	.998**	−.971**	−.999**	.995**	−.998**
	Sig. (bilateral)	.000	.000		.000	.000	.000	.000	.000	.000	.000	.000
	N	20	20	20	20	20	20	20	20	20	20	20
H_Al	Correlación de Pearson	.945**	.836**	.936**	1	−.939**	−.942**	.925**	−.921**	−.943**	.936**	−.938**
	Sig. (bilateral)	.000	.000	.000		.000	.000	.000	.000	.000	.000	.000
	N	20	20	20	20	20	20	20	20	20	20	20
MO	Correlación de Pearson	−.994**	−.941**	−.998**	−.939**	1	.995**	−.996**	.975**	.997**	−.991**	.996**
	Sig. (bilateral)	.000	.000	.000	.000		.000	.000	.000	.000	.000	.000
	N	20	20	20	20	20	20	20	20	20	20	20
N	Correlación de Pearson	−.992**	−.941**	−.997**	−.942**	.995**	1	−.994**	.967**	.997**	−.990**	.994**
	Sig. (bilateral)	.000	.000	.000	.000	.000		.000	.000	.000	.000	.000
	N	20	20	20	20	20	20	20	20	20	20	20
Na	Correlación de Pearson	.991**	.948**	.998**	.925**	−.996**	−.994**	1	−.960**	−.996**	.993**	−.996**
	Sig. (bilateral)	.000	.000	.000	.000	.000	.000		.000	.000	.000	.000
	N	20	20	20	20	20	20	20	20	20	20	20
K	Correlación de Pearson	−.978**	−.940**	−.971**	−.921**	.975**	.967**	−.960**	1	.974**	−.962**	.969**

Correlaciones												
		pHenH2O	pHenKCl	CE	H_Al	MO	N	Na	K	Ca	Mg	P
	Sig. (bilateral)	.000	.000	.000	.000	.000	.000	.000		.000	.000	.000
	N	20	20	20	20	20	20	20	20	20	20	20
Ca	Correlación de Pearson	−.996**	−.947**	−.999**	−.943**	.997**	.997**	−.996**	.974**	1	−.991**	.997**
	Sig. (bilateral)	.000	.000	.000	.000	.000	.000	.000	.000		.000	.000
	N	20	20	20	20	20	20	20	20	20	20	20
Mg	Correlación de Pearson	.991**	.947**	.995**	.936**	−.991**	−.990**	.993**	−.962**	−.991**	1	−.995**
	Sig. (bilateral)	.000	.000	.000	.000	.000	.000	.000	.000	.000		.000
	N	20	20	20	20	20	20	20	20	20	20	20
P	Correlación de Pearson	−.993**	−.941**	−.998**	−.938**	.996**	.994**	−.996**	.969**	.997**	−.995**	1
	Sig. (bilateral)	.000	.000	.000	.000	.000	.000	.000	.000	.000	.000	
	IM	20	20	20	20	20	20	20	20	20	20	20

***. La correlación es significativa en el nivel 0,01 (2 colas).*

Table 13.
Matrix of correlations of samples from Yamora and Viacha.

KMO and Bartlett's Test		
Kaiser Meyer Olkin Measure of Sampling Adequacy		.865
Bartlett's Test of Sphericity	Approx. Chi-Square	715.671
	df	55
	Sig.	.000

Table 14.
KMO and Bartlett's sphericity results for soils samples.

Total variance explained									
Component	Initial eigenvalues			Extraction sums of squared loadings			Rotation sums of squared loadings		
	Total	% of variance	Cumulative %	Total	% of variance	Cumulative %	Total	% of variance	Cumulative %
1	10.707	97.338	97.338	10.707	97.338	97.338	5.673	51.572	51.572
2	.166	1.511	98.848	.166	1.511	98.848	5.200	47.276	98.848
3	.063	.569	99.417						
4	.038	.342	99.759						
5	.011	.103	99.862						
6	.006	.056	99.918						
7	.004	.041	99.959						
8	.003	.024	99.983						
9	.001	.012	99.995						
10	.000	.003	99.998						
11	.000	.002	100.000						
Extraction Method: Principal Component Analysis.									

Table 15.
Extraction of principal components with a total explained variance of 98.84%.

Rotated component matrix[a]		
	Component	
	1	2
pHenH2O	.711	.699
pHenKCl	.876	.461
CE	.728	.683
H_Al	.476	.870
MO	−.717	−.694
N	−.710	−.699
Na	.739	.667
K	−.725	−.657

Rotated component matrix[a]		
	Component	
	1	2
Ca	−.718	−.694
Mg	.723	.682
P	−.717	−.693

Component score coefficient matrix		
	Component	
	1	2
pHenH2O	.011	.123
pHenKCl	1.195	−1.122
CE	.106	.024
H_Al	−1.184	1.367
MO	−.041	−.092
N	−.007	−.128
Na	.185	−.059
K	−.174	.049
Ca	−.047	−.086
Mg	.095	.035
P	−.045	−.087

Extraction Method: Principal Component Analysis.
Rotation Method: Varimax with Kaiser Normalization.
Component Scores.
[a]Rotation converged in 3 iterations.

Table 16.
Rotated component matrix and component score coefficient matrix for the Yamora and Viacha samples.

$$PC1 = 0.123 * pH(H2O) - 1.122 * pH(KCl) + 0.024 * CE + 1.367 * H_{Al} - 0.092 * OM - 0.128 * N - 0.059 * Na - 0.049 * K - 0.086 * Ca + 0.035 * Mg - 0.087 * P \quad (4)$$

The score values are the following (**Table 17**):

The representation of the components for the Yamora and Viacha samples are shown in **Figure 7**. For both PC1 (**Figure 7a**) and PC2 (**Figure 7b**), the positive values indicate that the pH parameters in KCl, Na^+, EC, Mg^{2+}, pH in H_2O, and H-Al overlap the parameters of N, OM, P, Ca^{2+}, K^+. This means that if the soils have positive values of PC1 and PC2, then the soil has high pH values, high Na^+ concentration, and high EC. On the other hand, if the soil has negative values of the components, then the soil is rich in OM, P, N, which represents a much more suitable land for agriculture.

In the case of the Yamora samples, its PC1 and PC2 is negative, therefore, this soil is rich in OM, P and N, which represents a much more suitable soil for agriculture. In the results for the Viacha samples, the PC1 and PC2 are positive, therefore, this soil is shown as a soil not so suitable for agriculture (**Figure 8**).

Location	CP_1	CP_2
Yamora	−0.6039	−0.7968
Yamora	−2.9276	1.6177
Yamora	−0.7392	−0.6686
Yamora	0.4893	−2.0047
Yamora	−0.3933	−0.9410
Yamora	−0.3570	−1.0346
Yamora	−1.5191	0.1140
Yamora	−0.4907	−0.8760
Yamora	−0.2710	−1.0577
Yamora	−0.2547	−1.0504
Viacha	0.8095	0.5245
Viacha	0.8482	0.5004
Viacha	0.4159	1.0341
Viacha	0.4046	1.0660
Viacha	0.9214	0.4989
Viacha	0.3742	1.0584
Viacha	0.1010	1.3822
Viacha	0.6979	0.5567
Viacha	1.2224	0.0200
Viacha	1.2719	0.0567

Table 17.
Principal component coefficients for the Yamora and Viacha samples.

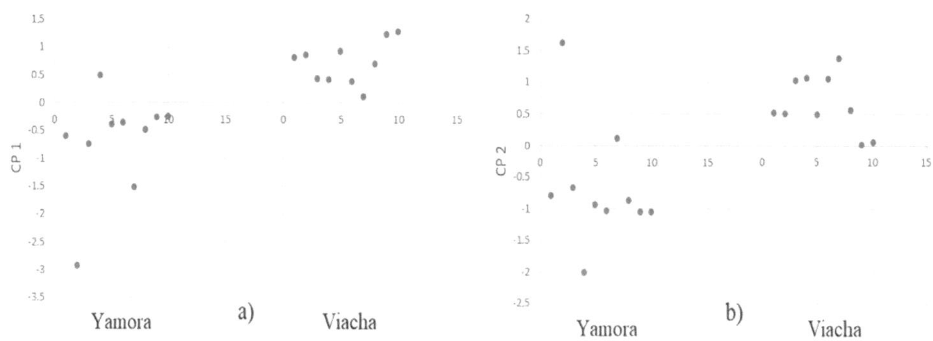

Figure 7.
Principal Component Analysis of the Yamora and Viacha samples a) Principal Component PC1, b) Principal Component PC2.

It can be observed that the main components accurately classify the two types of soils. In addition, a correlation can be observed for each type of soil (**Figure 9**).

The slope of both is approximately the same and the characterization of the soils is given by the ordinate to the origin (**Figure 9**). Soils with more suitable characteristics for cultivation, that is, the parameters N, OM, P, Ca^{2+}, K^+ overlap the pH in KCl, Na^+,

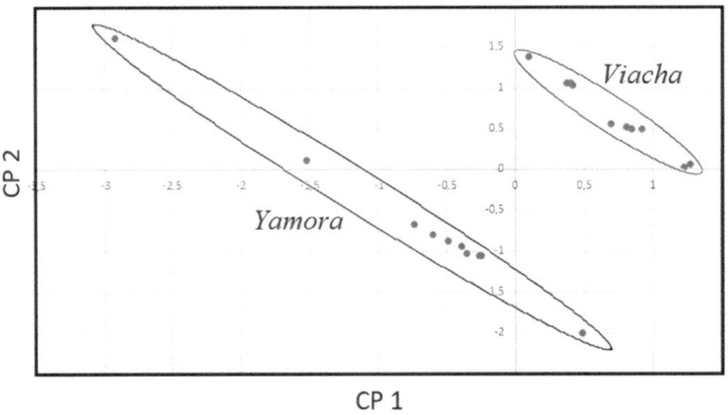

Figure 8.
Principal components PC1 and PC2 from Yamora and Viacha samples.

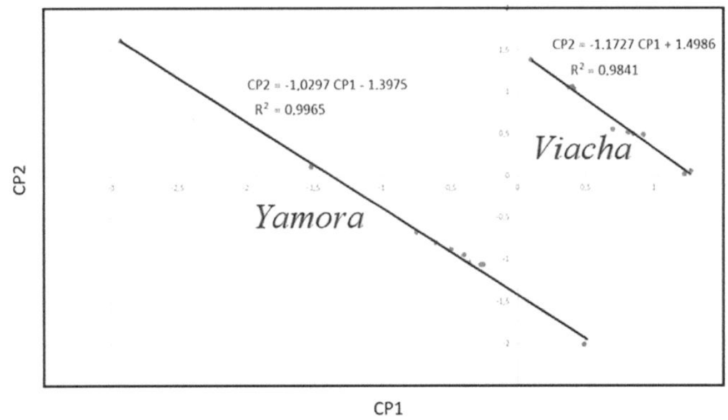

Figure 9.
The main components PC1 and PC2 show correlation for each type of soil, Yamora and Viacha.

EC, Mg^{2+}, pH in H_2O and H-Al tend towards smaller or even negative ordinates to the origin. In this case, the main components are capable of classifying and characterizing the soils with high precision. Thus, the multivariate analysis of soils constitutes an important tool for classifying soils.

It should be considered that the principal components give us a stand point in the data analysis. These must be complemented with other methods of multivariate analysis. In this case; for example, multivariate discriminant analysis can be applied [5].

The coefficients of the standardized canonical discriminant function indicate that the most appropriate parameters considered in the discriminant function are N, Na^+, K^+, Mg^{2+} and P. The parameters that are important to define soil fertility are: pH and OM. In addition, other factors that intervene in soil formation are the presence of minerals that contain exchange cations (Na^+, K^+, Mg^{2+} and Ca^{2+}), decreases in soil acidification and, the decomposition process of minerals.

The general discriminant function obtained for the two types of soils is [6]:

$$D = -18.418 + 118.391 * N - 8.267 * Na + 67.852 * K - 11.752 * Mg + 0.114 * P \quad (5)$$

While the discriminant functions by group are:

$$D_{Yamora} = -4971.556 + 10936.736 * N - 553.970 * Na + 5449.136 * K - 206.719 * Mg + 13.873 * P \quad (6)$$

$$D_{Viacha} = -2725.256 - 3502.174\, N + 454.280\, Na + 2826.109\, K + 1226.527\, Mg - 0.023\, P \quad (7)$$

The results of the application of the discriminant function in the classification of the samples in both places indicate that the 20 samples can be classified 100% correctly. Therefore, the application of these functions in the classification of new soil samples has a high probability of classifying them correctly. In this way, it is possible to classify the soils through five parameters and the discriminant function, and thus, determine its chemical fertility. This information can be complemented to the main components.

3. Conclusions

The data analysis by main Principal Components Analysis for the reduction of dimensions in data was applied to soil samples. It is shown that this tool is fundamental and fully applicable, since it allows the characterization and classification of soil samples with precision. This brings a better interpretation of the results.

Acknowledgements

Due acknowledgement to Springer Nature for giving permission to reproduce **Table 1** from "Sources and behavior of arsenic and trace elements in groundwater and surface water in the Poopó Lake basin, Bolivian Altiplano" by Oswaldo Eduardo Ramos Ramos, Luis Fernando Cáceres, Mauricio Rodolfo Ormachea Muñoz, Prosun Bhattacharya, Israel Quino, Jorge Quintanilla, Ondra Sracek, Roger Thunvik, Jochen Bundschuh, Maria Eugenia Garcia., Environmental Earth Sci., 66: 793 – 807, 2012.

A due acknowledgement to Revista Boliviana de Química for giving permission to reproduce the Table and the equation from "Chemometric evaluation of internal reference material (IRM) of agricultural soils in the two provincial municipalities of La Paz" by Rolando Mamani Quispe, Leonardo Guzmán Alegria, Jorge Chungara Castro, Oswaldo Eduardo Ramos Ramos, Revista Boliviana de Química, Vol. 39, No 4, pp 181 – 189, 2019; and "Análisis multivariable en la clasificación de suelos para la agricultura en el valle y Altiplano Boliviano" by Rolando Mamani Quispe, Oswaldo Eduardo Ramos Ramos, Jorge Chungara Castro, Leonardo Guzmán Alegría, Revista Boliviana de Química, Vol. 38, No 3, pp 126 – 132, 2021.

A due acknowledgment to José Antonio Bravo, Ph.D., Chief Editor of Revista Boliviana de Química for gramatical revision of the document.

Conflict of interest

The authors declare no conflict of interest.

Author details

Oswaldo Eduardo Ramos Ramos* and Leonardo Guzmán Alegría
Chemistry Department, Natural and Pure Sciences Faculty, Universidad Mayor de San Andrés (UMSA), La Paz, Bolivia

*Address all correspondence to: oramos@fcpn.edu.bo

References

[1] Martín Q., Cabero M. T., de Paz Y. 2007. Tratamiento estadístico de datos con SPSS, pg. 328, España, Universidad de Salamanca. Ed. Thomson

[2] Eduardo RRO, Fernando CL, Rodolfo OMM, Prosun B, Israel Q, Jorge Q, et al. Sources and behavior of arsenic and trace elements in groundwater and surface water in the Poopó Lake basin. Bolivian Altiplano. Environmental Earth Science. 2012;**66**: 793-807

[3] Eduardo Ramos Ramos Oswaldo. Geochemistry of trace elements in the Bolivian Altiplano – Effects of natural processes and anthropogenic activities. PhD Thesis, TRITA LWR PHD-2014:04. 2014

[4] Rolando MQ, Leonardo GA, Jorge CC, Eduardo RRO. Chemometric evaluation of internal reference material (IRM) of agricultural soils in the two provincial municipalities of La Paz. Revista Boliviana de Química. 2019; **39**(4):181-189

[5] Rolando MQ, Eduardo RRO, Jorge CC, Leonardo GA. Análisis multivariable en la clasificación de suelos para la agricultura en el valle y Altiplano Boliviano. Revista Boliviana de Química. 2021;**38**(3):126-132

[6] Mongay FC. Quimiometría. España: Universitat de Valencia; 2005. p. 245 Ed. PUV

Chapter 6

Unveiling Chromosome Changes Compatible with Climate Warming

Esteban Vegas, Lluís Serra, Ferran Reverter and Josep Maria Oller

Abstract

This work illustrates the use of multivariate descriptive statistics methods adapting different dimensional reduction techniques to the analysis of specific data. The particular nature of the data provides an opportunity to illustrate the pedagogical aim of this work. More explicitly, we will analyze and relate two different kinds of information: climate and genetic data and their change over time. We will show that the relation between both types of changes can be attributed to the unveiled genetic changes being compatible with the adaptation of *Drosophila subobscura* populations to warmer climates. The climate data are the monthly average temperatures of various populations in Europe and America at two different time periods separated by about 25 years. The genetic data include different profiles of *Drosophila subobscura* chromosomal polymorphisms that, as shown in the scientific literature, are related to the adaptation of the species to different climates. The genetic data have been obtained in the same populations and times as the climate data.

Keywords: dimensional reduction, interpretability, chromosomal polymorphism, cline, climate change

1. Introduction

Drosophila subobscura, a small species of fruit fly, has been an object of fascination for geneticists and ecologists due to its remarkable ability to adapt to changing environmental conditions. In recent years, as climate change accelerates and ecosystems undergo changes, researchers have focused their attention in understanding how this species responds to changing climate. It has five pairs of acrocentric chromosomes, all of which exhibit polymorphism for inversions. The frequencies of these chromosome arrangements show a clinal change according to latitude and, therefore, with climate [1, 2]. This particular fly species is native to the Old World and has a wide distribution in its native regions. In February 1978, *Drosophila subobscura*, which had never before been documented in the Americas, was discovered by chance in southern Chile [3]. The colonization of this species most likely originates from a Mediterranean population, although the precise origin remains undisclosed. A few years later (1982), it was also discovered in North America [4].

To forecast evolutionary responses to natural or anthropogenic perturbations, it is fundamental to determine whether evolutionary trajectories are predictable or idiosyncratic [5]. The predictability of evolution is evaluated by determining whether replicate populations show convergent responses, in particular, by collecting and analyzing genetic and climatological data over time.

Within the field of data analysis, multivariate statistical analysis [6, 7] serves as an indispensable instrument to achieve a deep understanding of data. Specifically, dimension reduction techniques can provide a more complete and holistic view of the data being analyzed. There are a large number of dimension reduction techniques [8, 9], although the principal component analysis (PCA) [9, 10] is the most widely used dimension reduction technique in research. In our case, we will use the power of PCA, matrix algebra operations, and statistics optimization to combine genetic and climatological information at two moments in time. With the aim of obtaining a joint representation that helps interpret whether populations show convergent responses with respect to genetic and climatological data in both Europe and America.

2. Methods and results

Both population genetics data (from *Drosophila subobscura*) and climate data will be considered. First, we shall summarize the workflow of the data analysis carried out as follows:

Figure 1.
European populations and their relationship with respect to the NE to SW cline. The data can be seen in ***Table 1****.*

1. To find a convenient Euclidean space to represent genetic data, starting from a reasonable standard distance between genetic profiles.

2. Obtain an interpretable direction in this Euclidean space, in terms of a geographic cline according to the results of a previous paper [11] using only European populations, see **Figure 1** and **Table 1**.

3. Once this direction has been established, we shall find another direction orthogonal to the first, with all the data, carrying out a standard PCA but onto the orthogonal complement space of the previously mentioned first direction, obtaining a 2D representation (**Figure 2**). The old and new data from the same geographic population have been represented by two points (one black dot and one red dot, respectively) joined by a *time arrow*, which visualizes the genetic profile changes.

4. With the climate data, that is, mean month temperatures corresponding to the same populations and periods, we make a PCA BIPLOT analysis, interpreting the first two components as *warm* weather and *extreme* (inter-seasonal) weather, obtaining also a 2D representation (**Figure 3**). The old and new temperature data from the same geographic population are also represented as two points joined by a *time arrow*, which visualizes the mean temperature changes.

5. Finally, we integrate both types of information representing the first and second principal components of climate data analysis into the 2D genetic space representation (**Figure 4**).

2.1 Genetic data

We dispose of data of the chromosomal inversion frequencies corresponding to the five chromosomes of *Drosophila subobscura* in several European and American

Population	Longitude/latitude	NE to SW in °	First component
Groningen	53°13′N–6°35′E	0.00°	−1.914748
Wien	48°13′N–16°22′E	0.03°	−1.952958
Louvain-la-neuve	50°43′N–4°37′E	3.20°	−1.429999
Tübingen	48°32′N–9°04′E	3.24°	−1.749663
Leuk	46°19′N–7°39′E	5.85°	−0.981962
Villars	45°26′N–0°44′E	9.93°	−0.743784
Montpellier	43°36′N–3°53′E	10.08°	−0.766615
Lagrasse	43°05′N–2°37′E	11.18°	−0.427634
Queralbs	42°13′N–2°10′E	12.18°	0.611528
Calvià	39°33′N–2°29′E	14.35°	1.513612
Riba-roja	39°33′N–0°34′W	15.96°	1.407313
Málaga	36°43′N–4°25′W	20.37°	1.909295
Punta Umbría	37°10′N–6°57′W	21.54°	2.208420

Table 1.
*Longitude and latitude of European geographic populations. Also given are degrees with respect to the NE and SW direction. See also **Figure 1**. The correlation with the first principal component obtained with the 13 European geographic populations at the first time period is $r = 0.9628$.*

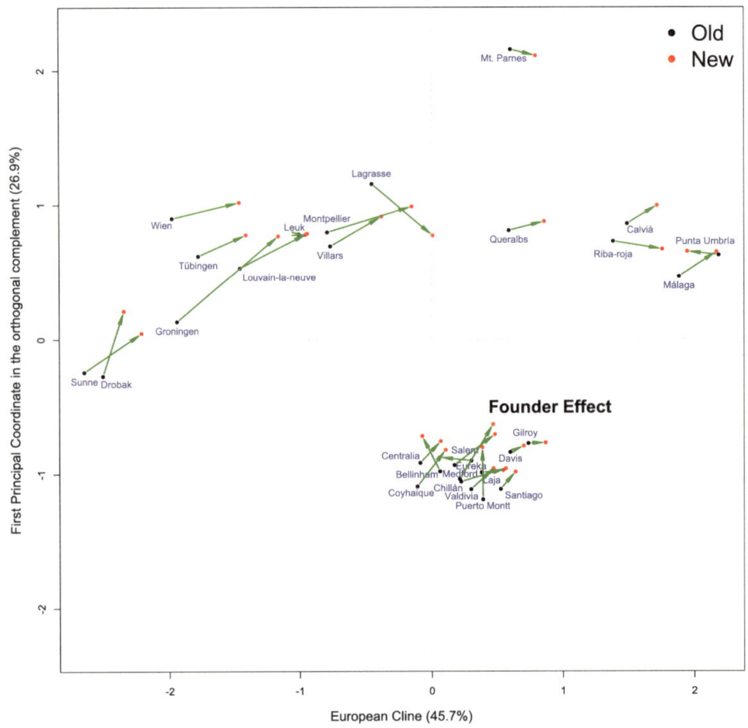

Figure 2.
2D representation of the abstract genetic space with the 29 geographic populations studied in two periods. The genetic profile changes are visualized by the arrows. The populations of the old period of time studied are represented with black dots and, in the new period of time, in red dots.

populations, each of them sampled in the same geographical places but at two different periods widely separated in the time, 24 years on average. Thus, although we have a total of 29 *geographic* populations, we shall consider a total of $N = 29 \times 2 = 58$ *statistical* populations (29 *old* and 29 *new*). Let $k = 5$ be the number of chromosomes analyzed (A, J, U, E, and O) and $m_1, \ldots, m_k$ the number of different chromosomal arrangements located on chromosome A, J, U, E, and O, respectively. Then, given a statistical population P_α, we denote by $\left(p_{\alpha i1}, \ldots, p_{\alpha i m_i}\right)$ a vector whose components are the relative frequencies of each chromosomal arrangement located on the i chromosome at population P_α, with $p_{\alpha ij} \geq 0$ where $\sum_{j=1}^{m_i} p_{\alpha ij} = 1$ with $i = 1, \ldots, k$ and $\alpha = 1, \ldots, N$. Further details on the data may be found in [11, 12].

Thus, each geographic population at a given time is characterized as point P in a *Genetic Space*, P, a point determined by

$$P = \left(p_{11}, \ldots, p_{1m_1}, p_{21}, \ldots, p_{2m_2}, \ldots, p_{k1}, \ldots, p_{km_k}\right) \quad (1)$$

From a mathematical point of view, this space is a *manifold with boundary* of dimension $n = \sum_{i=1}^{k} m_i - k$, with coordinate system defined through (1). In the present study, $\sum_{i=1}^{k} m_i = 50$, and $n = 45$.

If we consider a discrete multivariate Bernoulli distribution corresponding to each chromosome and we assume independence between them, it is well known that the Fisher information matrix induces a Riemannian structure [13] in the considered

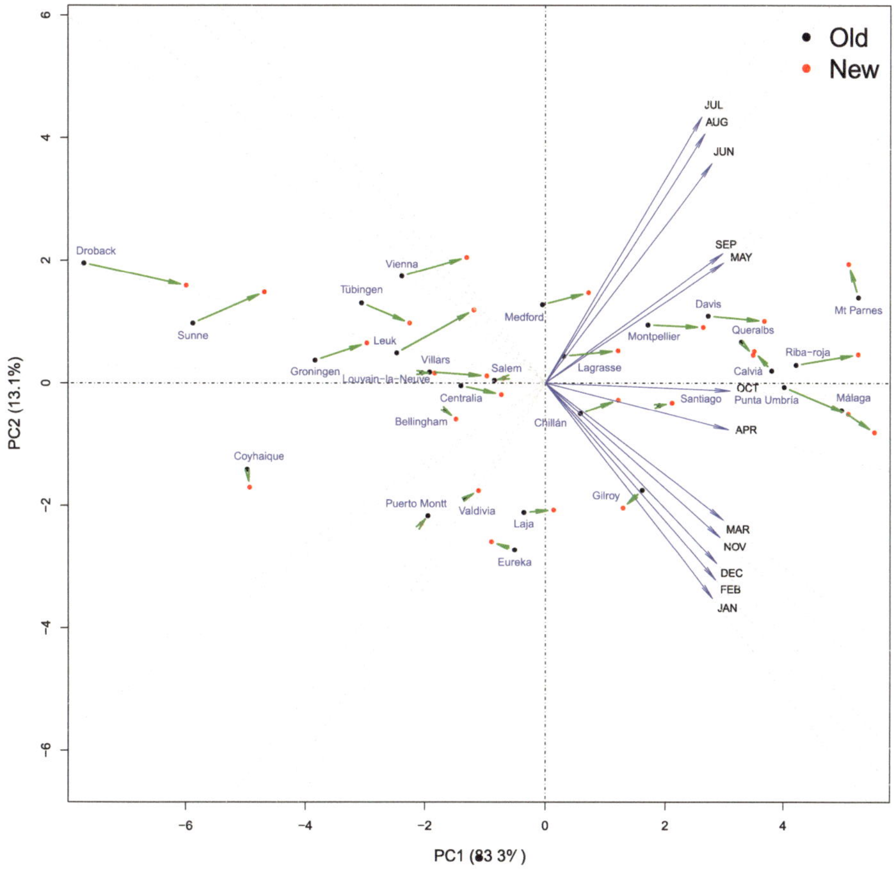

Figure 3.
2D BIPLOT with climate data. The 29 geographic populations are studied in two periods. The temperature changes are visualized by the arrows. The populations of the old period of time studied are represented with black dots, and the new period of time are in red dots. Each one of the variables, mean month standardized temperatures, is represented as blue straight lines whose direction vectors are the gradient of each variable (up to a proportionality constant: Scaled for legibility reasons) and therefore indicates its maximum increase by unit length. The horizontal axis may be interpreted as warm climate (increasing to the right) and extreme interseasonal climate (increasing upward).

manifold that represents the genetic space whose distance will be given in terms of the square root of the sum of squares of the information metric between multivariate Bernoulli distribution corresponding to each chromosome, these being equal to the Bhattacharyya distance, up to at most a multiplicative factor, that is, the distance between two points P_α and P_β of coordinates determined by $\left(p_{\alpha 11}, \dots, p_{\alpha k m_k}\right)$ and $\left(p_{\beta 11}, \dots, p_{\beta k m_k}\right)$, respectively, will be

$$d(P_\alpha, P_\beta) = 2\sqrt{\sum_{i=1}^{k}\left(\arccos\left(\sum_{j=1}^{m_i}\sqrt{p_{\alpha ij}p_{\beta ij}}\right)\right)^2} \qquad \alpha, \beta = 1, \dots, N \tag{2}$$

Although we do not know the *exact coordinates* of each statistical population, we have available *reasonable* estimations of them that, for the sake of simplicity, we shall denote in the same way $\left(p_{\alpha 11}, \dots, p_{\alpha k m_k}\right)$ and $\left(p_{\beta 11}, \dots, p_{\beta k m_k}\right)$, respectively.

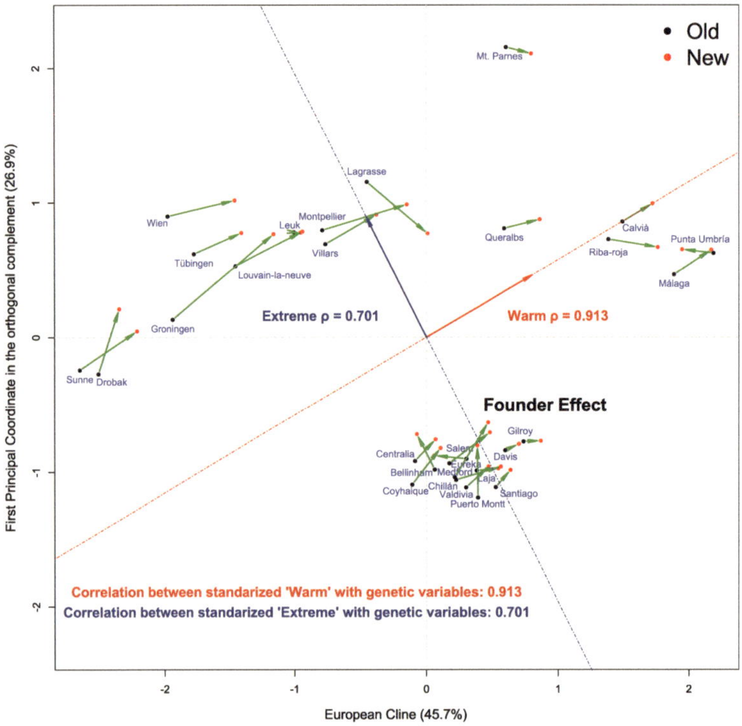

Figure 4.
2D dynamic diagram genetic space and climate. Note that changes in the genetic profile are compatible with an adaptation to a warm climate.

Therefore, from (2), we can obtain a $N \times N$ square matrix $\boldsymbol{D} = \left(d\left(P_\alpha, P_\beta\right)\right)$, whose entries are the *estimated* distances between all pairs of the $N = 58$ studied statistical populations. This distance, up to a multiplicative factor, was already used in [11], with a subset of the data analyzed in the present paper corresponding to 13 geographic European populations in two different time periods ($N_E = 13 \times 2 = 26$ statistical populations).

From this distance matrix, we can carry out a principal coordinate analysis (PCoA) [14, 15]. Specifically, if we let $\mathbf{1}_N$ be a $N \times 1$ vector whose entries are all equal to 1, $\boldsymbol{I}_N$ the $N \times N$ identity matrix $\boldsymbol{H} = \boldsymbol{I}_N - \frac{1}{N}\mathbf{1}_N\mathbf{1}_N{}^t$, where the symbol ^{t} indicates the transpose vector or matrix, the $N \times N$ centering matrix $\boldsymbol{A} = \frac{1}{2}\ \boldsymbol{D}\circ\boldsymbol{D}$, where $\circ$ denotes the Hadamard product, we can obtain a spectral decomposition in matrix form $\boldsymbol{A} = \boldsymbol{Q}\Lambda\boldsymbol{Q}^t$ where Λ is a $N \times N$ diagonal matrix with the ordered eigenvalues of $\boldsymbol{A}$ $\lambda_1 \geq \lambda_2 \geq, \ldots, \geq \lambda_N$ and $\boldsymbol{Q}$ is a $N \times N$ ortonormal matrix with the corresponding normalized eigenvectors. The eigenvalues obtained range from $\lambda_1 = 84.86376$ to $\lambda_N = -0.2118237$, being strictly positive the 38 first eigenvalues. Furthermore, $\sum_{i=1}^{38}\lambda_i = 176.8488$, whereas $\sum_{i=39}^{58}\lambda_i = -0.5911$. The positive part is approximately the 99.67% of the addition of the absolute value of all eigenvalues. This fact suggests the possibility to map each one of the $N = 58$ statistical populations into a point in $\mathbb{R}^q$ where $q = 38$ with coordinates given by the 38 first principal coordinates. If $\tilde{\boldsymbol{Q}}$ is the $N \times q$ matrix with the first 38 columns of $\boldsymbol{Q}$ and $\tilde{\Lambda}$ is the $q \times q$ diagonal matrix of the strictly positive ordered eigenvalues $\lambda_1 \geq \ldots, \geq \lambda_q > 0$, the 38 principal coordinates of the statistical populations are the rows of the $N \times q$ matrix $\boldsymbol{X}$ given by

$$X = \tilde{Q}\tilde{\Lambda}^{1/2} \tag{3}$$

The ordinary Euclidean distance between the rows of X almost reproduces the original distance given in (2). In other words, if we identify each statistical population P_α, with $\alpha = 1, \dots, N$ as a point that has $q = 38$ Euclidean coordinates $(x_{\alpha 1}, \dots, x_{\alpha q})$ such that the ordinary Euclidean distance among them,

$$d_E(P_\alpha, P_\beta) = \sqrt{\sum_{i=1}^{q} (x_{\alpha i} - x_{\beta i})^2} \qquad \alpha, \beta = 1, \dots, N \tag{4}$$

is *very similar* to $d(P_\alpha, P_\beta)$. Moreover, we can build the interdistance Euclidean matrix $N \times N$, $\boldsymbol{D}_E = (d_E(P_\alpha, P_\beta))$, obtaining the following measures:

• Kruskal Stress [15, 16]

$$Stress(\boldsymbol{D}, \boldsymbol{D}_E) = \sqrt{\frac{\sum_{i,j=1}^{N} (d(P_\alpha, P_\beta) - d_E(P_\alpha, P_\beta))^2}{\sum_{i,j=1}^{q} d(P_\alpha, P_\beta)^2}} = 0.0030 \tag{5}$$

this small Kruskal Stress indicates that $\boldsymbol{D}$ and $\boldsymbol{D}_E$ are almost equal.

• Another measure that suggests a similar behavior of both distances is the Pearson correlation between the entries of both matrices,

$$\rho(\boldsymbol{D}, \boldsymbol{D}_E) = 0.99999 \tag{6}$$

• And the following measure, similar to a *relative error* of the approximation, gives

$$Rel.\ Error(\boldsymbol{D}, \boldsymbol{D}_E) = \frac{\sum_{i,j=1}^{N} |d(P_\alpha, P_\beta) - d_E(P_\alpha, P_\beta)|}{\sum_{i,j=1}^{q} d(P_\alpha, P_\beta)} = 0.00248 \tag{7}$$

this small relative error also indicates that $\boldsymbol{D}$ and $\boldsymbol{D}_E$ are almost equal.

All these measures indicate a great similarity between (2) and (4) and suggest using, for simplicity, the representation of the $N = 58$ statistical populations in $\mathbb{R}^q$, with the ordinary Euclidean distance as the abstract genetic space into which we represent all of our statistical populations because their proximity relations using the Euclidean distance (4) are almost the same that the ones obtained with (2) proposed initially. The sum of squares between this distance of all considered point will be

$$SD_E^2 = \sum_{\alpha=1}^{N} \sum_{\beta=1}^{N} d_E^2(P_\alpha, P_\beta) = 2N\ \operatorname{tr}(\boldsymbol{X}^t \boldsymbol{H} \boldsymbol{X}) \tag{8}$$

The next step is to obtain a good two-dimensional Euclidean representation that helps us interpret the observed genetic variability. To achieve this purpose, it is important to maximize the variability explained in the 2D graphic output compared with the total variability in the whole Euclidean space ($q = 38$ dimensions). This could be obtained by performing a simple PCA from the data matrix $\boldsymbol{X}$ [8–10, 15–18].

We also look for *easily interpretable directions* in this abstract Euclidean genetic space. Specifically, we will obtain an interpretable direction of this space that will be used as the first axis of the 2D graphic output, that is to say, we will try to interpret a direction of such space in terms of the northeast-southwest (***NE-SW***) geographic cline described in [11]. In that paper, this cline was identified as the first component of a principal coordinate analysis realized using (2), up to a proportionality constant, and obtained from the data corresponding with the 13 first European populations considered in the present chapter in the two time periods considered.

To obtain in our abstract genetic space the direction linked with the geographic cline described in [11], we shall find the first principal component obtained from the first $13 \times 2 = 26$ rows of data matrix $\boldsymbol{X}$, which correspond to the European populations studied in [11]. This component will be linked to the geographic cline because (2) and (4) are very similar. Specifically, let $\boldsymbol{X}_\varepsilon$ be the 26×38 matrix with the 26 first rows of $\boldsymbol{X}$, and $\boldsymbol{X}_\rho$ be the 32×38 matrix with the remaining rows of $\boldsymbol{X}$, that is, $\boldsymbol{X}^t = \left(\boldsymbol{X}_\varepsilon^t, \boldsymbol{X}_\rho^t\right)$. The direction searched will be given by the first principal component obtained from $\boldsymbol{X}_\varepsilon$. To obtain a PCA with these data, with the same basic notation as before, if we let $\mathbf{1}_\varepsilon$ be a $r \times 1$ vector whose entries are all equal to 1, with $r = 26$, $\boldsymbol{I}_\varepsilon$ the $r \times r$ identity matrix, and $\boldsymbol{H}_\varepsilon = \boldsymbol{I}_\varepsilon - \frac{1}{r}\mathbf{1}_\varepsilon\mathbf{1}_\varepsilon^t$ where the symbol ^{t} indicates the transpose vector or matrix, the $r \times r$ centering matrix, we can build the covariance matrix corresponding to the *abstract* data matrix $\boldsymbol{X}_\varepsilon$, a 38×38 matrix given by $\boldsymbol{S}_\varepsilon = \boldsymbol{X}_\varepsilon^t\boldsymbol{H}_\varepsilon\boldsymbol{X}_\varepsilon/r$. Then, we will diagonalize this matrix; and because we have $r = 26$ points in an Euclidean space, there will be a maximum of 25 strictly positive covariance matrix eigenvalues, $\lambda_{\varepsilon 1} \geq \lambda_{\varepsilon 2} \geq \ldots \geq \lambda_{\varepsilon p} > 0$, obtaining a spectral decomposition whose matrix form may be expressed as $\boldsymbol{S}_\varepsilon = \boldsymbol{U}_\varepsilon\boldsymbol{D}_\varepsilon\boldsymbol{U}_\varepsilon^t$, where $\boldsymbol{D}_\varepsilon$ is a diagonal $p \times p$ matrix with the obtained ordered positive eigenvalues, and $\boldsymbol{U}_\varepsilon$ is a $q \times p$ matrix whose columns are normalized eigenvector corresponding to the aforementioned positive eigenvalues. The first principal component, given by the first column of $\boldsymbol{U}_\varepsilon$, referred hereafter simply as the 38×1 column vector $\boldsymbol{u}$, it will be the direction in the abstract q-dimensional genetic space linked with the geographic cline. This direction summarizes the $100\lambda_{\varepsilon 1}/\mathrm{tr}(\boldsymbol{S}_\varepsilon) = 69.3\%$ variability of the statistical populations corresponding to European populations. The correlation between the first component obtained and the first principal coordinate analysis described in [11] corresponding to the old data is -0.999999, which indicates the same direction. The negative sign is simply due to the fact that we choose the sign of the first component obtained to be positively correlated with the geographic cline oriented from the ***NE*** to ***SW***, see **Table 1**.

To clearly show the relationship between the first component and the northeast to southwest geographic cline, we can correlate a measurement in degrees along the ***NE*** to ***SW*** cline, say $\boldsymbol{NE/SW}$, from 0.00° of Groningen to 21.54° of Punta Umbría, obtained through the standard latitude and longitude coordinates of the different populations, with the values of the first principal component of each geographic population at the old time period studied, say $\boldsymbol{Y}_{\varepsilon old}$, obtaining a correlation $\rho(\boldsymbol{NE/SW}, \boldsymbol{Y}_{\varepsilon old}) = 0.9628$, see **Table 1**.

Next, to obtain the second direction of the abstract genetic space that we will use to obtain the 2D representation, we will project all 58 points that represent each statistical population onto the orthogonal complement space of the span of $\mathbf{u}$, $\langle\mathbf{u}\rangle^\perp$. On this 37-dimensional space, we will carry out a standard PCA and will use the first principal component, say the $q \times 1$ column vector $\mathbf{v}$, orthogonal to $\mathbf{u}$ and unitary, on the projected points on $\langle\mathbf{u}\rangle^\perp$ as the second direction of the space where we will project all the points, namely, the span of $\mathbf{u}$ and $\mathbf{v}$, $\langle\mathbf{u}, \mathbf{v}\rangle$. Specifically, the projection onto $\langle\mathbf{u}\rangle^\perp$ is realized by means of the product $\mathbf{X}(\mathbf{I} - \mathbf{u}\mathbf{u}^t)$, where $\mathbf{I}$ is the 38×38 identity matrix;

this $N \times q$ matrix with the coordinates of the $N = 58$ populations in $\langle \mathbf{u} \rangle^{\perp}$ is a centered matrix, and its covariance will be $\mathbf{S} = (\mathbf{I} - \mathbf{uu}^t)\mathbf{X}^t\mathbf{X}(\mathbf{I} - \mathbf{uu}^t)/N$.

Then, we will diagonalize this matrix **S** and will have a maximum of $q = 38$ strictly positive covariance matrix eigenvalues, $\lambda_1 \geq \lambda_2 \geq \ldots \geq \lambda_q \geq 0$, obtaining a spectral decomposition whose matrix form can be expressed as $\mathbf{S} = \mathbf{V}\Lambda\mathbf{V}^t$, where Λ is a diagonal $q \times q$ matrix with the obtained ordered positive eigenvalues, and **V** is a $q \times q$ matrix whose columns are normalized orthogonal eigenvector corresponding to the aforementioned eigenvalues. The first principal component will be given by the first column of **V**, referred hereafter simply as the $q \times 1$ column vector **v**; it will define the direction in the abstract q-dimensional genetic space that we shall use to complete the desired 2D representation. The direction determined by **v** summarizes the $100\lambda_1/\mathrm{tr}(\mathbf{S}) = 49.2\%$ variability of the statistical populations corresponding to all populations in $\langle \mathbf{u} \rangle^{\perp}$. The 2D final plot will be obtained with the projection of the rows of **X** into the span of **u**, **v**, namely, $\langle \mathbf{u}, \mathbf{v} \rangle$. Introducing the $q \times 2$ matrix $\mathbf{P} = (\mathbf{u}, \mathbf{v})$, that is, a two column matrix whose columns are the column vectors **u** and **v**, the coordinates of the N statistical populations will be rows of the $N \times 2$ matrix $\mathbf{Y} = \left(y_{\alpha j}\right)$ given by

$$\mathbf{Y} = \mathbf{XP} \tag{9}$$

We can easily quantify the percentage of variability represented by each one of the 2D-plot axes, computing first the sum of the squared distance between all the points if we project them in the first axis, SD_{a1}^2 and in the second one SD_{a2}^2 of the plot, that is

$$SD_{ai}^2 = \sum_{\alpha=1}^{N}\sum_{\beta=1}^{N}\left(y_{\alpha i} - y_{\beta i}\right)^2 \qquad \text{with} \qquad i = 1, 2 \tag{10}$$

and comparing these quantities with the total variability SD_E^2 given in (8), we obtain the percentages θ_{ai} given by $\theta_{ai} = 100 \times SD_{ai}^2/SD_E^2$ with $i = 1, 2$, which in the present case are equal to $\theta_{a1} = 45.5\%$ and $\theta_{a2} = 26.8\%$, thus, the variability explained by both axis $\theta_{a1} + \theta_{a2}$ will be equal to 72.3%. We can see the described 2D representation in **Figure 2**; the coordinates of the populations are given by the Eq. (9).

To interpret the change of the genetic profile of the 29 geographic populations between the two periods studied, separated in time on average in just over 24 years, it will be convenient to define two 29×1 column vectors $\mathbf{Y}_1$ and $\mathbf{Y}_2$ that contains the two coordinates of each one of the 29 populations in the plot, the populations located in the same order in both vectors, at the old period $\mathbf{Y}_1$ and at the new one $\mathbf{Y}_2$. The change of the genetic profiles of each geographic population between periods may be visualized by $\boldsymbol{\delta} = \mathbf{Y}_2 - \mathbf{Y}_1$, given by the green arrows in **Figure 3** from the black points (populations of the old time period) to the red points (populations of the new time period) (**Table 2**).

Additionally, among the total of 29 geographic populations considered, in 25 of them in the course of the time between the first and the second period of time studied, a displacement occurs, increasing the value of the first variable used for the 2D representation, which is a direction that represents the **NE/SW** cline. Moreover, this low value is hardly compatible with a random change (in an ordinary bilateral sign test, $p-$ value$\approxeq 1.0 \times 10^{-4}$). Let us also observe the limited dispersion of American populations, as a consequence of the so-called *founder effect* produced by the recent

Population	δ_1	δ_2	Population	δ_1	δ_2
Montpellier	0.642373	0.190589	Lagrasse	0.459498	−0.381803
Queralbs	0.268763	0.066369	Riba-roja	0.374971	−0.059232
Calvià	0.228700	0.134521	Punta Umbría	**−0.237194**	0.027191
Málaga	0.282278	0.184082	Groningen	0.780543	0.635947
Louvain-la-neuve	0.513534	0.256845	Villars	0.389008	0.219267
Tübingen	0.364991	0.157145	Wien	0.516528	0.118356
Leuk	0.050395	−0.004103	Sunne	0.438114	0.288466
Drobak	0.158006	0.481837	Mt. Parnes	0.189012	−0.047603
Gilroy	0.130422	0.004212	Davis	0.101785	0.046262
Eureka	0.305773	0.228315	Medford	0.251930	0.405888
Salem	**−0.241229**	0.023833	Centralia	0.153696	0.160464
Bellinham	**−0.133972**	0.262271	Santiago	0.113615	0.127641
Chillán	0.339739	0.097727	Laja	0.164784	0.015877
Valdivia	0.168553	0.155098	Puerto Montt	**−0.005071**	0.386463
Coyhaique	0.212375	0.270840	–	–	–

The few negative values in the second column are highlighted in contrast to the majority of the positive values.

Table 2.
The components of vectors $\boldsymbol{\delta} = (\boldsymbol{\delta}_1, \boldsymbol{\delta}_2)$ represent the profile genetic changes in each of the 29 studied geographic populations. Observe that the 29 components of $\boldsymbol{\delta}_1$, the first column vector of $\boldsymbol{\delta}$, are positive in 25 of the 29 cases (each one corresponding to a different population). The shift is compatible with profiles more adapted to the climate of **SW**.

colonization of *Drosophila subobscura* to the New World in the late 70s of the last century when this species was accidentally introduced to America. This effect refers to the reduction of genomic variability due to a small group of individuals separating from the original population [12].

2.2 Climate data

Along with the data analysis of chromosomal polymorphisms in *Drosophila subobscura*, we recorded meteorological data for the 4 years immediately preceding each biological sample from the nearest meteorological station for each population using NASA GISS. (http://data.giss.nasa.gov/gistemp/) and NOAA (http://www.ncdc.noaa.gov/oa/climate/climatedata.html). Then, we calculate the average monthly temperatures during these 4 years, in degrees Celsius, seasonally adjusting them and taking into account the hemisphere of origin. Thus, to each of the statistical populations considered ($N = 58 = 29 \times 2$), we will associate a vector in $\mathbb{R}^{12}$ whose components are precisely the average temperature values per month, using standardized data to make the analysis invariant under linear changes. Then, we perform an ordinary PCA. Specifically, let $\mathbf{W} = (w_{\alpha j})$ be a 58×12 real matrix whose entries are the month mean values attached to each statistical population determined by the 29 different geographic sites and two periods, from January to December, seasonally adjusted by hemisphere (for instance, January is the real January in the northern hemisphere but July in the southern hemisphere and similar with the other months) and already standardized. Then, with the same basic notation as before, we can compute the corresponding covariance matrix equal to the correlation matrix because the data are standardized, given by $\mathbf{R_W} = \mathbf{W}^t\mathbf{HW}/N$. Then, we will diagonalize this matrix and obtain 12 ordered nonnegative eigenvalues $\lambda_1 \geq \lambda_2 \geq \ldots \geq \lambda_{12} \geq 0$ obtaining a spectral decomposition whose matrix form may be expressed as $\mathbf{R_W} = \mathbf{T}\Gamma\mathbf{T}^t$, where Γ is a diagonal 12×12 matrix with the ordered nonnegative eigenvalues, and $\mathbf{T}$ is a 12×12 orthogonal matrix whose columns are normalized eigenvector corresponding to the aforementioned nonnegative eigenvalues. Because $\lambda_1 = 10.00127$, the first principal component given by the first column of $\mathbf{T}$ summarizes the 83.3% of mean temperature variance; whereas since $\lambda_2 = 1.56655$, the second principal component, given by the second column of $\mathbf{T}$ summarizes the 13.1% of mean temperature variance. Therefore, the two main principal components summarizes the 96.4% of mean temperature total variance. Explicitly, introducing the 12×2 matrix $\Upsilon = (\mathbf{t}_1, \mathbf{t}_2)$, where $\mathbf{t}_1$ and $\mathbf{t}_2$ are the first two columns of $\mathbf{T}$ and these two components on the $N = 58$ statistical populations will be given by the two columns of the $N \times 2$ matrix $\mathbf{C} = (\mathbf{c}_1, \mathbf{c}_2) = (c_{\alpha j})$ with $j = 1, 2$ given by

$$\mathbf{C} = \mathbf{W}\Upsilon \tag{11}$$

Classical PCA BIPLOT of climate data, see **Figure 3**, where the populations (labeled in blue) and climate variables (arrows in blue) are represented. The coordinates of the populations corresponding to the old temperature measurements appear in black and the coordinates corresponding to the new measurements (taken 4 years later as average) appear in red. It will be convenient to interpret the variables in order to compute the loadings of the principal components to each variable, and these quantities also obtain a standard BIPLOT [19, 20].

The different variables (from January to December) can be represented by different straight lines through the origin. **Table 3** can facilitate the interpretation of the principal components. Because the coefficients determining the first principal component are all positive, we can interpret this component as a size-related component. Specifically, associated with higher temperature, let us call *warm climate* the first component direction. The blue arrows show the displacement of the populations

Population/month	January	February	March	April	May	June
Droback old	−7.025	−8.850	−4.050	2.150	7.975	12.800
Málaga new	14.075	14.825	16.075	17.275	19.250	22.000
Vienna new	0.375	1.725	5.550	10.700	16.350	20.025
Eureka old	9.533	8.833	9.867	10.233	12.175	13.625
Population/month	**July**	**August**	**September**	**October**	**November**	**December**
Droback old	16.600	13.950	9.900	4.725	−1.200	−5.175
Málaga new	24.000	24.575	22.175	20.450	17.425	15.300
Vienna new	21.025	21.525	14.825	10.375	6.125	0.100
Eureka old	14.650	15.125	14.175	12.375	10.350	9.625

Table 3.
The mean temperatures in Celsius degrees of four statistical populations in the different month referred to as northern hemisphere seasons, that is, we shift the data corresponding to populations of the southern hemisphere since in that hemisphere the summer period is the winter in the northern hemisphere, among others.

induced by the time interval. We observe that the arrows are mostly aligned with the first component and, consequently, aligned with an increase of mean temperatures: a shift to a Warmer climate. We can interpret the second principal component as a component related to interseasonal extreme weather, since the loadings of the warmer month are positives and negative the loadings of colder months let us call *extreme climate* the second component direction.

2.3 Climate data into the 2D genetic space representation

We can represent the first and second principal component of the previously mentioned climate data PCA into the 2D graphic obtained using only genetic data.

One way to proceed is to find the linear combination of the genetic variables $(\mathbf{Y}_1,\mathbf{Y}_2)$ that have the maximum correlation with the first and second principal components of climate data $\mathbf{c}_1$ and $\mathbf{c}_2$, that is, find the coefficients $\boldsymbol{\alpha}^t = (\alpha_1, \alpha_2)$ and $\boldsymbol{\beta}^t = (\beta_1, \beta_2)$ such that

$$\underset{\boldsymbol{\alpha}^t=(\alpha_1,\alpha_2)}{\arg\max}\, cor\left(\mathbf{c}_1, \alpha_1\mathbf{y}_1 + \alpha_2\mathbf{y}_2\right) \quad \underset{\boldsymbol{\beta}^t=(\beta_1,\beta_2)}{\arg\max}\, cor\left(\mathbf{c}_2, \beta_1\mathbf{y}_1 + \beta_2\mathbf{y}_2\right) \tag{12}$$

The direction defined by $\boldsymbol{\alpha}^t = (\alpha_1, \alpha_2)$ is the linear combination of the abstract genetic variables $\mathbf{Y}_1$ and $\mathbf{Y}_2$ that fit the most with the first temperature component $\mathbf{c}_1$. In a similar way, the direction defined by $\boldsymbol{\beta}^t = (\beta_1, \beta_2)$ is the linear combination of the abstract genetic variables $\mathbf{Y}_1$ and $\mathbf{Y}_2$ that best fits with the second temperature component $\mathbf{c}_2$.

After some straightforward computation, the sought coefficients $\boldsymbol{\alpha}$ and $\boldsymbol{\beta}$ that define the previously mentioned directions in the abstract genetic space and that imply unit variance of the vectors $\mathbf{Y}\boldsymbol{\alpha}$ and $\mathbf{Y}\boldsymbol{\beta}$ are given by

$$\boldsymbol{\alpha} = \frac{\left(\mathbf{Y}^t\mathbf{Y}\right)^{-1}\mathbf{Y}^t\mathbf{c}_1}{\sqrt{\mathbf{c}_1^t\mathbf{Y}\left(\mathbf{Y}^t\mathbf{Y}\right)^{-1}\mathbf{Y}^t\mathbf{c}_1}} \quad \text{and} \quad \boldsymbol{\beta} = \frac{\left(\mathbf{Y}^t\mathbf{Y}\right)^{-1}\mathbf{Y}^t\mathbf{c}_2}{\sqrt{\mathbf{c}_2^t\mathbf{Y}\left(\mathbf{Y}^t\mathbf{Y}\right)^{-1}\mathbf{Y}^t\mathbf{c}_2}} \tag{13}$$

and the achieved maximum correlation with $\mathbf{c}_1$ and $\mathbf{c}_2$ are respectively

$$\rho_1 = \frac{\sqrt{\mathbf{c}_1{}^t\mathbf{Y}\left(\mathbf{Y}^t\mathbf{Y}\right)^{-1}\mathbf{Y}^t\mathbf{c}_1}}{\sqrt{\mathbf{c}_1{}^t\mathbf{c}_1}} \quad \text{and} \quad \rho_2 = \frac{\sqrt{\mathbf{c}_2{}^t\mathbf{Y}\left(\mathbf{Y}^t\mathbf{Y}\right)^{-1}\mathbf{Y}^t\mathbf{c}_2}}{\sqrt{\mathbf{c}_2{}^t\mathbf{c}_2}} \tag{14}$$

In the present example, we obtain $\rho_1 = 0.913466$ that shows that the direction corresponding to the first component of the PCA in **Figure 4** are closely linearly related with *warm climate*, whereas the value obtained for $\rho_2 = 0.700762$ shows an important although weaker linear relationship of the direction corresponding to the second principal component in **Figure 4**, interpreted as *extreme climate*.

Furthermore, to study the relationship of the variable which *represents* the *Warm climate* with the observed shifts of the genetic population profiles of each geographic population at different times, given by $\boldsymbol{\delta}$, the green arrows from the black points (populations of the old time period) to the red points (populations of the new time period), we can compute the scalar product of these vectors $\boldsymbol{\delta}$ with the variable $\mathbf{Y}\boldsymbol{\alpha}$. Among the total of 29 geographic populations considered, in 27 (all with the exception

Population	$\langle \delta_i, \mathbf{Y}\alpha \rangle$	$\langle \delta_i, \mathbf{Y}\beta \rangle$	Population	$\langle \delta_i, \mathbf{Y}\alpha \rangle$	$\langle \delta_i, \mathbf{Y}\beta \rangle$
Montpellier	0.601249	−0.121418	Lagrasse	0.191069	−0.548328
Queralbs	0.245389	−0.062709	Riba-roja	0.272323	−0.222716
Calvià	0.244812	0.016150	Punta Umbría	**−0.177003**	0.131730
Málaga	0.310490	0.036006	Groningen	0.917106	0.212613
Louvain-la-neuve	0.528855	−0.004008	Villars	0.412010	0.018966
Tübingen	0.364161	−0.025477	Wien	0.467364	−0.128710
Leuk	0.038379	−0.026496	Sunne	0.483173	0.058340
Drobak	0.348535	0.357530	Mt. Parnes	0.129084	−0.128070
Gilroy	0.106166	−0.055364	Davis	0.102679	−0.004933
Eureka	0.349669	0.064752	Medford	0.388557	0.247314
Salem	**−0.181777**	0.130567	Centralia	0.196843	0.073252
Bellinham	0.013925	0.294317	Santiago	0.149672	0.062186
Chillán	0.316573	−0.066951	Laja	0.139006	−0.060550
Valdivia	0.206240	0.061739	Puerto Montt	0.174221	0.346502
Coyhaique	0.294650	0.144961	–	–	–

The few negative values in the second column are highlighted in contrast to the majority of the positive values.

Table 4.
The second and fifth columns show the scalar product between δ_i and $\mathbf{Y}\alpha$. Analogously, the third and sixth columns provide the scalar product between δ_i and $\mathbf{Y}\beta$, where $\mathbf{Y}\alpha$ is the direction that represents warm climate in the 2D representation of the genetic spaces and $\mathbf{Y}\beta$ is the direction which represents extreme climate in the 2D representation of the genetic space. Observe that the scalar product between δ_i and $\mathbf{Y}\alpha$ is positive in almost all population, which is compatible with a profile shift to more warm adapted genetic configurations.

of Punta Umbría and Salem) of them this scalar product is positive indicating that the shift is compatible with a genetic adaptation to a warmer climate. Moreover, this low value is hardly compatible with a random change (in an ordinary bilateral sign test, $p - \text{value} \approxeq 1.6 \times 10^{-6}$) (**Table 4**).

3. Conclusions

In this work, we illustrate a procedure that combines dimension reduction for temporal data and the integration of multiple multivariate sources, in our case, genetic data and climate data. We have proposed a formulation to enrich dimension reduction by choosing interpretable directions in the representation space rather than maximizing variability.

Analysis of the genetic data consistently reveals a shift of populations along the European northeast to southwest cline when comparing the two time periods. On the other hand, the analysis of climate data reveals a first component associated with warm climate. At the same time, a shift of populations in the direction of higher temperature is observed when comparing time points. When we integrate the climatic with the genetic information, we have verified that both displacements, corresponding to the populations in the genetic space and the populations in the climate space, respectively, are highly aligned, suggesting that all populations along the cline are evolving toward the adaptation to a warm climate.

We emphasize that the methodology used in this study is very versatile and easily applicable for multivariate data integration in other areas of application.

Acknowledgements

This research was funded by the Ministerio de Ciencia e Innovación (Spain) (Ref: PID2019-104830RB-I00/AEI DOI: 10.13039/501100011033).

Conflict of interest

The authors declare no conflict of interest.

Nomenclature

PCA	principal component analysis
PCoA	principal coordinate analysis
2D ...	2 dimensional ...
NE	Northeast
SW	Southwest
NASA	National Aeronautics and Space Administration
GISS	Goddard Institute for Space Studies
NOAA	National Oceanic and Atmospheric Administration

Author details

Esteban Vegas[1*†], Lluís Serra[2†], Ferran Reverter[1†] and Josep Maria Oller[1†]

1 Faculty of Biology, Department of Genetics, Microbiology and Statistics, University of Barcelona, Barcelona, Spain

2 University of Barcelona, Barcelona, Spain

*Address all correspondence to: evegas@ub.edu

† These authors contributed equally.

References

[1] Prevosti A. Chromosomal polymorphism in *Drosophila subobscura* populations from Barcelona (Spain). Genetics Research. 1964;**5**(1):27-38

[2] Menozzi P, Krimbas CB. The inversion polymorphism of D. Subobscura revisited: Synthetic maps of gene arrangement frequencies and their interpretation. Journal of Evolutionary Biology. 1992;**5**(4):625-641

[3] Ayala FJ, Serra L, Prevosti A. A grand experiment in evolution: The *Drosophila subobscura* colonization of the Americas. Genome. 1989;**31**(1): 246-255

[4] Beckenbach AT, Prevosti A. Colonization of North America by the European species, *Drosophila subobscura* and *D. ambigua*. American Midland Naturalist. 1986;**115**(1):10-18

[5] Huey RB, Gilchrist GW, Carlson ML, Berrigan D, Serra L. Rapid evolution of a geographic cline in size in an introduced fly. Science. 2000;**287**(5451):308-309

[6] Johnson RA, Wichern DW. Applied Multivariate Statistical Analysis. 6th ed. Upper Saddle River, NJ: Pearson; 2007

[7] Izenman AJ. Modern Multivariate Statistical Techniques. Vol. 1. New York: Springer; 2008

[8] Nanga S, Bawah A, Acquaye B, Billa M, Baeta F, Odai N, et al. Review of dimension reduction methods. Journal of Data Analysis and Information Processing. 2021;**9**:189-231. DOI: 10.4236/jdaip.2021.93013 [Accessed: September 12, 2023]

[9] Burges CJ. Dimension reduction: A guided tour. Foundations and Trends® in Machine Learning. 2010;**2**(4):275-365

[10] Hotelling H. Analysis of a complex of statistical variables into principal components. Journal of Educational Psychology. 1933;**24**(6):417-441. DOI: 10.1037/h0071325 [Accessed: September 12, 2023]

[11] Balanyà J, Solé E, Oller JM, Sperlich D, Serra L. Long-term changes in the chromosomal inversion polymorphism of *Drosophila subobscura*. II. European populations. Journal of Zoological Systematics and Evolutionary Research. 2004;**42**:191-201. DOI: 10.1111/j.1439-0469.2004.00274.x [Accessed: September 12, 2023]

[12] Balanyá J, Oller JM, Huey RB, Gilchrist GW, Serra L. Global genetic change tracks global climate warming in *Drosophila subobscura*. Science. 2006; **313**(5794):1773-1775. DOI: 10.1126/science.1131002 [Accessed: September 12, 2023]

[13] Burbea J, Rao CR. Entropy differential metric, distance and divergence measures in probability spaces: A unified approach. Journal of Multivariate Analysis. 1982;**12**(4): 575-596

[14] Gower JC. Some distance properties of latent root and vector methods used in multivariate analysis. Biometrika Trust. 1966;**53**:325-338. DOI: 10.1093/biomet/53.3-4.325 [Accessed: September 12, 2023]

[15] Borg I, Groenen PJ. Modern Multidimensional Scaling: Theory and Applications. New York, US: Springer Science & Business Media; 2005

[16] Kruskal JB. Multidimensional scaling by optimizing goodness of fit to a nonmetric hypothesis. Psychometrika. 1964;**29**(1):1-27

[17] Karl Pearson FRS. LIII. On lines and planes of closest fit to systems of points in space. The London, Edinburgh, and Dublin Philosophical Magazine and Journal of Science. 1901;**2**(11):559-572. DOI: 10.1080/14786440109462720 [Accessed: September 12, 2023]

[18] Radhakrishna RC. The use and interpretation of principal component analysis in applied research. Sankhyā: The Indian Journal of Statistics, Series A. 1964;**26**(4):329-358 http://www.jstor.org/stable/25049339 [Accessed: September 12, 2023]

[19] Gabriel KR. The Biplot graphic display of matrices with application to principal component analysis. Biometrika. 1971;**58**:453-467. DOI: 10.1093/biomet/58.3.453 [Accessed: September 24, 2023]

[20] Gower, John C, Lubbe SG, Roux NJL. Understanding Biplots. Hoboken, NJ: John Wiley & Sons; 2011

Chapter 7

Multivariate Analysis of Cranial Measurements of Cameroon's Blue Duiker (*Cephalophus monticola*)

Miantsia Fokam Olivier, Felix Meutchieye and Evaristus Tsi Angwafo

Abstract

The blue duiker (*Cephalophus monticola*) is exclusively an African wild Bovidae. It is a principal source of protein in the African forest zones and contributes to the nutrition of local populations. The methodology used is the opportunistic method which consists of taking the heads of blue duikers from carcasses encountered randomly and opportunistically in villages, urban markets, and checkpoints. Thus, 60 skulls were collected and measured from January to December 2018. Descriptive statistics and multivariate analysis were done using SPSS version 21.0 software and XLSTAT-Pro version 7.5.2 software. The measures of the skulls (60 in total and in mm) point out that: the total length (117.36 ± 3.51; 118.23 ± 4.38 and 118.47 ± 4.09), the length of row of cheek teeth (35.35 ± 2.88; 36.39 ± 3.82 and 36.28 ± 3.67) the zygomatic arc height (10.40 ± 1.50; 11.06 ± 1.12 and 11.17 ± 1.10) in these three areas respectively indicate a significant difference (<0.05). The Principal Component Analysis (PCA) enables us to see the level of genetic variabilities of blue duiker through skull measurements. These variable measurements are close together from one to another where there is a high similarity between species. Grouping these biometric characteristics permitted us to identify three structures of the blue duiker, corresponding to the three sub-species found in Cameroon.

Keywords: characterization, skull, blue duikers, biometry, Cameroon

1. Introduction

Global biodiversity is currently undergoing an unprecedented crisis [1, 2], to the point that some scientists are now talking about a sixth extinction [3]. Species extinction could be up to 1000 times higher than last century [4]. Some authors such as [5] argue that the consequences of these extinctions are comparable to the effects of climate change in terms of ecosystem change. This phenomenon is amplified by human actions on biodiversity, including commercial hunting, which is perceived as one of the major threats to this biodiversity. Three species alone account almost 70% of the bushmeat sold in all the markets of the country of the COMIFAC countries: the blue duiker, *Cephalophus monticola*; the African brush-tailed porcupine, *Atherurus africanus* and the Greater spot-nosed monkey, *Campithecus nictitans*; [6]. The blue duiker account 39% of this harvest, as the main supplier of bushmeat [7, 8]. This animal, commonly called as "hare", is well known to forest populations for its food use. Beyond that, it could have other, little-known uses. Market observations of the carcasses of these animals show a variation in the color of the fur, ranging from dark grey to light grey to bright black. These variations in fur levels could lead to errors of assessment on the part of the population living near forests, on the one hand, and on the other hand, among urban consumers and the entire control chain.

Beyond the food discolouration and in the context of the "biodiversity crisis", the characteristics of the blue duiker remain insufficient. However, it is the first approach for the identification and sustainable use of the species [9]. The first step in this characterization of the blue duiker is based on knowledge of the variations of the biometric features [9]. These are found on the live animal and on the skull. However, most of the phenotypes of the majority of natural animal species are not recorded, such as those of the blue duiker [10]. Furthermore, due to the lack of comprehensive information on population structure and geographical distribution, many animal populations in developing regions are commonly considered "indigenous" or "traditional". Therefore, there is a need to establish simplified and consistent phenotypic characterization procedures to help countries conduct a more comprehensive inventory of their animal genetic resources [11].

At a time when the countries of the Congo Basin, including Cameroon, have begun the process of developing bushmeat management strategies, poor identification of blue duiker can have serious consequences on the conservation of this species whose meat is highly valued in the forest zone [12–14]. Studies of the blue duiker in Gabon and Cameroon have examined its relationship to the structure of the environment [15], its diet [16], its place among primary consumers, particularly the frugivorous [17] and its abundance [18, 19]. Relatively maneuverable and easy to capture by hunters, the blue duiker is a suitable study for better craniometric characterization [20], although according to Dubost [21], in reality, it is not easy to characterize it.

Identifying the blue duiker is essential and the foundation of any successful modern management. In recent years, the need to properly identify an animal, to trace it through the production chain and ultimately in food products, that is to say, to have traceability, has become essential in order to record the Evolution of its weight gain, fertility, susceptibility to diseases, and thus to facilitate the selection and management of genetic resources.

2. Methodology

2.1 Study area

Cameroon is located in the west of the Central Sub Region of Africa, stretching from the Gulf of Guinea to Lake Chad. It falls between latitude 2° to 13° North of the equator and longitude 8° 30′ to 16° 10′ East of the Greenwich Meridian. The country covers a surface area of 475,385 km^2 and has a coastline of 402 km. It is bounded to the South by the Republic of Congo, Gabon, Equatorial Guinea and the Atlantic Ocean, to the west by the Republic of Nigeria, to the North by Lake Chad and to the East by the Republic of Chad and the Central African Republic. Blue duikers are in three agro-ecological zones of Cameroon (**Figure 1**): Western Highlands (WH) equals zone 3 falls between latitude 4°54′ to 6°36′ North and longitude 9°18′ to 11°24′ East; Monomodal Rain Forest (MRF) equals zone 4 falls between latitude 2°6′ to 6°12′ North and

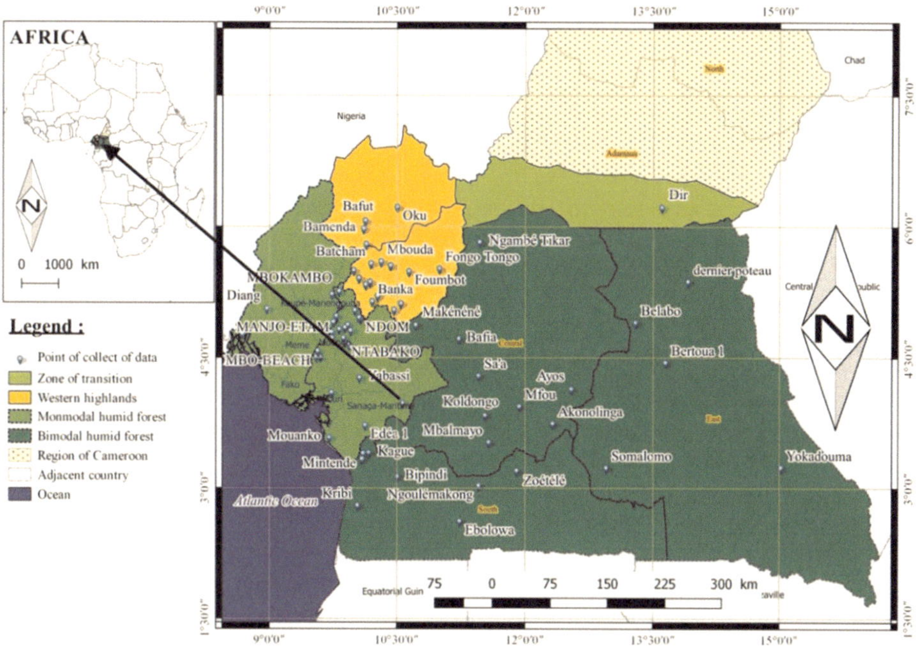

Figure 1
Study area of blue duiker's skull.

longitude is 8°48′ to 10°30′ East and Bimodal Rain Forest (BRF) equals zone 5 falls between latitude 10°30′ North to longitude 16°12′ East [22].

2.2 Data collection

The head is detached from the body by hunters or vendors using a machete. Once the head has been obtained, the skull is stripped of its skin and as much adherent flesh as possible using a very sharp knife. The lower jaw is separated and the tongue and eyes are removed; the cervical cavity is emptied of its contents. The skull and the lower jaw are emaciated as best as possible, they are immersed in a container filled with cold water for 1 or 2 days, renewing the water several times. The purpose of this operation is to purge all the blood vessels and soften the flesh in order to facilitate the rest of the work.

The skull and lower jaw are then boiled in a suitably sized container (washing pot) filled with water, taking care that the liquid still covers the skull. This cooking takes about 15–20 min.

We take out the skull and the jawbone and we strip off as much flesh and cartilage as possible with a sharp knife, pliers, a wire bent into a hook. The cooking vessel is rinsed and cleaned and the skull and jaw are returned to clean water. Boil again for a few minutes. Usually, this second boil is enough for all meat and cartilage debris to come off easily at the tip of the knife.

After checking that there are no more meaty particles left, especially in the cervical cavity, in the nasal cavities and under the millstones, the skull is rinsed with a jet of water, brushed well with an ordinary foaming detergent, rinsed again and air dried for 24 h. After these different operations, the different measurements are carried out [23]. Thus, a total of 14 cranial measurements and mass were taken. These measurements numbered from 1 to 10 are identical to those used by [24] and mainly used to identify wolf subspecies (*Canis lupus*). Measurements of 11, 12, 13, 14 and 16 quickly distinguish a wolf from a coyote, a wolf or a coyote from a dog as well as the whole range of their hybrids [24]. Therefore, the 15 measures retained consider that ruminants do not have canines. Thus, 60 skulls were measured at a rate of 20 (10 males and 10 females) per agro-ecological zone. According to [25], to perform craniometric measurements, five instruments are required: the vernier, the ruler, the compass, the bevel and the protractor. For our study, two instruments were used: the electronic vernier and the protractor. When using this type of vernier, it is important to check from time to time if the dial still indicates zero when the jaws are closed, otherwise, the yellow ABS button is pressed to reset to zero. The protractor is used exclusively to measure the orbital angle. To facilitate the reading of the latter, it is preferable to obtain a protractor of good size graduated to the nearest 0.5°.

The cranial measurement uses many existing possibilities [23]. The measurements enumerated from 1 to 10 are the same used by [25] and served especially for the classification of subspecies of wolf (*Canis lupus*). Measurements 11; 12; 13; 14 and 16 are the main ones to distinguish the Wolf of the Coyote [24]. For instance, ruminants do not have the canine, we have taken 14 measurements. Then, 60 skulls were measured haphazardly by the opportunistic method developed by [26], to evaluate the quantity of bush meat in the Central African villages. These 14 measures (in mm) have been taken by the Vernier with the skull mass (in g): total skull length, zygomatic width, jugal teeth, line length, palate maximal width, minimal palate width, postorbital apophyses width, the height between first molar and orbit, arcade zygomatic height, the fourth upper pre-molar length, second molar width, distance between the margin of the incisive row and the edge of the temporal condyle, condyle basal length, pre-maxillary width, orbiter angle and mass.

2.3 Statistical analysis

On the base of the 15 craniometrical measures, we have used statistical description. The Principal Component Analysis (PCA) has been carried out to evaluate the genetic variability of the blue duiker population studies [27]. The data analysis method is called the multivariate analysis, which consists of the transformation of the correlated variable in the new decorrelated variables from one another. These new variables are named "principal components", or principal axes. It enables us to reduce the number of variables and to send the least redundant information. The PCA enabled us to identify the least number of components or axes with better explained data variability. It consists in compressing the whole number of random variables, the first axes of the PCA are the best choices in terms of the inertia or of the variance.

In the construction of a phylogenetic tree or dendrogram following the protocol of Hierarchy Ascending Classification (HAC), we have used the Pearson correlation to identify the genetic type and the relationship that exists between each other [28]. It is one of the statistical methods of data analysis that enabled us to divide a whole number of data into different homogenize groups. In this case, the data of each subwhole number divides the common measurement, which often corresponds to the proximity criteria (similarities or dissimilarities) that we have defined to introduce the distance measurement and classes between objects. This technique of data analysis also enables us to hierarchy data thereby, to construct dendrograms that give evidence of the distance between groups or their similarities and dissimilarities.

Analysis of population structure has been realized by means of Factorial Discriminant Analysis (FDA) on the base of 15 measurements [27] thereby identifying the characters that distinguished better blue duikers as described. It enables us to represent graphically the different cranial measurements of blue duikers and the centre of their groups using more discriminant axes.

The software analysis SPSS version 21.0 and XLSTAT-Pro version 7.5.2. was used for data analysis.

3. Results

The measurements of the skull will be of two types: the measurements (total length; zygomatic width; jugal teeth row length; maximum palate width; minimum palate width; post orbital apophyses width; orbital first molar base height; zygomatic arch height; maximum fourth premolar length upper; maximum second molar width; palate length; Condylo-basal length; pre-maxillary width; orbital angle) and the mass. These measurements are shown in **Tables 1–12** according to agro-ecological zones and sex.

3.1 Blue duiker skull measurements

3.1.1 According to agro-ecological zones

Tables 1–5 show that the agro-ecological zone has no influence on the blue duiker skull measurements. So, there is no significant difference between these skull

Agro-ecological zones	**n**		**Skull length (mm)**		**Zygomatic width (mm)**	
		CV (%)	$\mu \pm \sigma$	CV (%)	$\mu \pm \sigma$	CV (%)
WH Z3	20	36.25	117.36 ± 3.51	2.99	55.45 ± 1.17	2.10
MRF Z4	20	29.45	118.23 ± 4.38	3.71	56.33 ± 1.34	2.38
BRF Z5	20	34.72	118.47 ± 4.09	3.45	56.30 ± 1.27	2.25
Total	**60**					
P-value			**0.000**		**0.000**	
P < 0.05.						

Table 1.
Total skull length, zygomatic width of skull of blue duiker.

Agro-ecological Zones	**n**	**Jugal teeth line length (mm)**		**Palate maximal width (mm)**		**Palate minimal width (mm)**	
		$\mu \pm \sigma$	CV (%)	$\mu \pm \sigma$	CV (%)	$\mu \pm \sigma$	CV (%)
WH Z3	20	35.35 ± 2.88	8.15	37.57 ± 2.74	7.30	20.19 ± 2.24	11.10
MRF Z4	20	36.39 ± 3.82	10.50	44.11 ± 19.10	43.29	21.03 ± 1.37	6.52
BRF Z5	20	36.28 ± 3.67	10.13	38.63 ± 3.45	8.94	21.19 ± 1.71	8.05
Total	**60**						
P-value		**0.000**		**0.000**		**0.000**	
P < 0.05.							

Table 2.
Jugal teeth line length, palate maximal width and minimal palate width of skull of blue duiker.

Agro-ecological Zones	n	Post-orbital apophyses width (mm)		Height between first molar and orbit (mm)		Arcade zygomatic height (mm)	
		μ ± σ	CV (%)	μ ± σ	CV (%)	μ ± σ	CV (%)
WH Z3	20	29.18 ± 6.26	21.45	14.30 ± 1.51	10.53	10.40 ± 1.50	14.46
MRF Z4	20	29.99 ± 7.25	24.16	15.10 ± 1.14	7.57	11.06 ± 1.12	10.13
BRF Z5	20	30.25 ± 6.80	22.49	15.23 ± 1.23	8.08	11.17 ± 1.10	9.83
Total	**60**						
P-value		**0.000**		**0.000**		**0.000**	

$P < 0.05$.

Table 3.
Post-orbital apophyses width, height between the first molar and orbit, arcade zygomatic height of skull of blue duiker.

Agro-ecological zones	n	Fourth upper pre-molar length (mm)		Second molar width (mm)		Length palate (mm)	
		μ ± σ	CV (%)	μ ± σ	CV (%)	μ ± σ	CV (%)
WH Z3	20	5.33 ± 0.86	16.18	4.87 ± 0.94	19.21	60.25 ± 5.38	8.93
MRF Z4	20	5.94 ± 1.50	25.31	5.55 ± 1.29	23.32	61.08 ± 6.37	10.43
BRF Z5	20	5.88 ± 1.35	23.03	5.58 ± 1.22	21.79	61.44 ± 6.02	9.80
Total	**60**						
P-value		**0.000**		**0.000**		**0.000**	

$P < 0.05$.

Table 4.
Fourth upper pre-molar length, second molar width, length palate of skull of blue duiker.

Agro-ecological Zones	n	Condyle basal length (mm)		Pre-maxillary width (mm)		Orbiter angle (°)	
		μ ± σ	CV (%)	μ ± σ	CV (%)	μ ± σ	CV (%)
WH Z3	20	100.21 ± 4.41	4.40	23.82 ± 1.22	5.12	27.75 ± 2.36	8.50
MRF Z4	20	101.22 ± 5.60	5.53	24.40 ± 1.48	6.07	28.25 ± 1.94	6.88
BRF Z5	20	101.52 ± 5.23	5.15	34.63 ± 5.84	16.87	28.35 ± 2.18	7.70
Total	**60**						
P-value		**0.000**		**0.000**		**0.000**	

$P < 0.05$.

Table 5.
Condyle basal length, pre-maxillary width, orbiter angle of skull of blue duiker.

measurements in the three agro-ecological zones. Therefore, the blue duiker population would be the same in the three agro-ecological zones. However, the territories of this animal have split over time. This division would be due to human activities, namely urbanization, agro-industry, road construction, logging, etc. The blue duiker has been able to adapt to its increasingly changing environment, with a notable growth in commercial hunting [28]. The sustainable management of this animal will be the grantee of cultural conservation and biodiversity in Cameroon.

3.1.2 According to the sex

Table 6 shows the P-value given for an equal two-sided test (0.340) above the 5% threshold. This means that the means of the zygomatic width measurements are significantly different between the male and female blue duiker. On the other hand, the P-value (0.000) of the total length of the skull is below the threshold of 5%. This means that the averages of total skull length are not significantly different and therefore influenced by sex.

Table 7 shows that the P-value given for an equal two-sided test (0.000) is below the 5% threshold. This means that the averages of the measurements of the row length of the cheek teeth, the maximum palate width and the minimum palatal width are not

Sex	n		Skull length (mm)		Zygomatic width (mm)	
		CV (%)	μ ± σ	CV (%)	μ ± σ	CV (%)
M	30	40.04	114.32 ± 0.82	0.72	55.67 ± 0.69	1.24
F	30	28.42	121.72 ± 1.77	1.45	56.38 ± 1.65	2.93
Total	**60**					
P-value			**0.000**		**0.340**	
T-test			−20.77		−21.17	
P < 0.05.						

Table 6.
Total skull length, zygomatic width of skull of blue duiker according to the sex.

Sex	n	Jugal teeth line length (mm)		Palate maximal width (mm)		Palate minimal width (mm)	
		μ ± σ	CV (%)	μ ± σ	CV (%)	μ ± σ	CV (%)
M	30	32.96 ± 0.62	1.89	35.24 ± 0.75	2.13	22.09 ± 0.80	3.63
F	30	39.06 ± 2.17	5.55	44.49 ± 14.80	32.92	19.52 ± 1.66	8.51
Total	**60**						
P-value		**0.000**		**0.000**		**0.000**	
T-test		−14.82		−3.59		7.64	
P < 0.05.							

Table 7.
Jugal teeth line length, palate maximal width and minimal palate width of skull of blue duiker according to the sex.

significantly different between the male and female blue duiker, therefore, the sex has an influence on these cranial measurements.

Table 8 shows that the P-value given for an equal two-sided test (0.000; 0.046) is below the 5% threshold. This means that the averages of the post-orbital process width measurements, the first orbital molar base height and the zygomatic arch height are not significantly different between the male and female blue duiker. Therefore, the sex has an influence on these cranial measurements.

Table 9 shows that the P-value given for an equal two-sided test (0.000; 0.002) is below the 5% threshold. This means that the averages of the measurements of the maximum fourth upper premolar length, the maximum second molar width and the palatal length are not significantly different between the male and female blue duiker, therefore, the sex has an influence on these cranial measurements.

Table 10 shows that the P-value given for an equal two-sided test (0.000; 0.026) is below the 5% threshold. This means that the means of the measurements of the condylo-basal length, the pre-maxillary width and the orbital angle are not significantly different between the male and female blue duiker, therefore, the sex has an influence on these cranial measurements.

Sex	n	Post-orbital apophyses width (mm)		Height between first molar and orbit (mm)		Arcade zygomatic height (mm)	
		$\mu \pm \sigma$	CV (%)	$\mu \pm \sigma$	CV (%)	$\mu \pm \sigma$	CV (%)
M	30	23.35 ± 1.11	4.76	15.22 ± 0.69	4.50	11.59 ± 0.40	3.44
F	30	36.27 ± 1.79	4.95	14.53 ± 1.72	11.86	10.16 ± 1.45	14.32
Total	**60**						
P-value		**0.000**		**0.046**		**0.000**	
T-test		**−33.540**		**2.044**		**5.202**	
P < 0.05.							

Table 8.
Post-orbital apophyses width, height between the first molar and orbit, arcade zygomatic height of skull of blue duiker according to the sex.

Sex	n	Fourth upper pre-molar length (mm)		Second molar width (mm)		Length palate (mm)	
		$\mu \pm \sigma$	CV (%)	$\mu \pm \sigma$	CV (%)	$\mu \pm \sigma$	CV (%)
M	30	5.04 ± 0.40	7.88	4.88 ± 0.38	7.87	55.32 ± 1.17	2.11
F	30	6.39 ± 1.49	23.33	5.79 ± 1.51	26.13	66.53 ± 1.86	2.80
Total	**60**						
P-value		**0.000**		**0.002**		**0.000**	
T-test		**−4.790**		**−3.183**		**−27.947**	
P < 0.05.							

Table 9.
Fourth upper pre-molar length, second molar width, length palate of skull of blue duiker according to the sex.

Sex	n	Condylo basal length (mm)		Pre-maxillary (mm)		Orbiter angle (°)	
		μ ± σ	CV (%)	μ ± σ	CV (%)	μ ± σ	CV (%)
M	30	96.54 ± 1.26	1.31	23.83 ± 1.29	5.39	30.00 ± 0.64	2.14
F	30	105.43 ± 3.06	2.90	31.41 ± 6.74	21.46	26.23 ± 1.28	4.87
Total	**60**						
P-value		**0.000**		**0.026**		**0.000**	
T-test		**−14.727**		**−1.125**		**14.419**	

P < 0.05.

Table 10.
Condyle basal length, pre-maxillary width, orbiter angle of skull of blue duiker according to the sex.

3.2 Blue duiker skull Mass

The mass of the skull of the blue duiker does not vary from one agro-ecological zone to another regardless of sex (**Table 11**).

It appears from **Table 11** that the agro-ecological zone has no influence on the mass of the blue duiker skull because $P < 0.05$. So, there is no significant difference between the mass of blue duikers in the three agro-ecological zones.

Table 12 shows the P-value given for an equal two-sided test (0.992) above the 5% threshold. This means that the averages of the cranial mass measurements are significantly different between the male and female blue duiker.

3.3 Principal Component Analysis (PCA) of cranial measurement of blue duiker's

The Principal Component Analysis was done to show the contribution of 15 cranial measurements to the explanation of genetic variabilities within the blue duiker's population. These 15 measurements have enabled us to obtain 15 proper values which permitted the construction of a correlation circle.

Principal component analysis (PCA) was carried out to show the contribution of 15 cranial measurements to the explanation of the total genetic variability observed within the blue duiker population. The eigenvalues and the factors are sorted in descending order of variability represented in **Table 13**.

Agro-ecological zones	n	Mass (g)	
		μ ± σ	CV (%)
WH Z3	20	6.49 ± 2.35	36.25
MRF Z4	20	7.59 ± 2.24	29.45
BRF Z5	20	8.24 ± 2.86	34.72
Total	**60**		
P-value		**0.000**	

P < 0.05.

Table 11.
Blue duiker skull mass according to the agro-ecological zones.

Sex	n	Mass (g)	
		$\mu \pm \sigma$	CV (%)
M	30	7.44 ± 2.98	40.04
F	30	7.44 ± 2.11	28.42
Total	**60**		
P-value		**0.992**	
T-test		**−0.010**	
$P < 0.05$.			

Table 12.
Blue duiker skull mass according to the sex.

In **Table 13**, the first eigenvalue emerges with a value of 7.53 and represents 50.22% of the variability. This means that if we represent the data on a single axis, then we will always have 50.22% of the total variability which will be preserved. Thus, we can deduce from the graph below that the pairs of variables (total length of the skull; mass), (zygomatic width; pre-maxillary width), (length of cheek teeth row; mass), (length of cheek teeth row; maximum palate width), maximum palate width; orbital angle), (zygomatic arch height; pre-maxillary width), (upper fourth premolar maximum length; pre-maxillary width) and (orbital angle; mass) show a significant correlation at the 5% level. However, not all cranial measurements are influenced by agro-ecological zones, which is not the case with sex, which influences these

Component	Initial eigenvalues			Extraction Sums of the squares of the factors retained		
	Total	% of variance	Cumulative %	Total	% of variance	Cumulative %
1	7.534	50.224	50.224	7.534	50.224	50.224
2	5.144	34.291	84.515	5.144	34.291	84.515
3	0.931	6.206	90.721			
4	0.738	4.922	95.643			
5	0.390	2.598	98.242			
6	0.110	0.731	98.973			
7	0.063	0.421	99.394			
8	0.042	0.279	99.673			
9	0.019	0.127	99.800			
10	0.012	0.081	99.881			
11	0.010	0.068	99.949			
12	0.005	0.033	99.981			
13	0.002	0.014	99.995			
14	0.000	0.003	99.998			
15	0.000	0.002	100.000			

Table 13.
Eigenvalues and cumulative proportion of principal components of cranial measurements.

measurements. We can conclude that there is a correlation between the different biometric characteristics at the 5% threshold.

The observation of **Table 13** shows that the orbital angle, the height of the zygomatic arches and the width of the minimum palate are orthogonal to each other, so they are significantly uncorrelated. On the other hand, the other measurements are close to each other, so they are significantly and positively correlated. The measurements of the skulls of blue duiker populations are mostly close to each other and therefore indicate how these measurements are grouped by agro-ecological zone according to their similarities (**Figure 2**).

Figure 2 makes it possible to represent the cranial measurements on a two-dimensional map, and thus to identify trends. We see that, on the basis of the biometric variables available, they are grouped according to the axes F1 (50.22%) and F2 (34.29%). We see in this figure that on the basis of the biometric variables available, the characteristics grouped according to the three agro-ecological zones are quite specific. We notice that the zones of BRF Z2, WH Z3, of MRF Z4 have measurements that are specific to them. These measurements are completely isolated and well away from the centre of the mark. The cranial measurements of blue duikers are mostly closer one against the other, indicating that these measurements construct a similar structure.

3.4 Hierarchy Ascending Classification (HAC) of blue duikers of Cameroon

The dendrogram in **Figure 3** illustrates the relationship between the three subclasses of cranial measurements based on similarity. The HAC will then gather the individuals iteratively in order to produce a dendrogram. That is, to identify subclasses of observations with similar measurements.

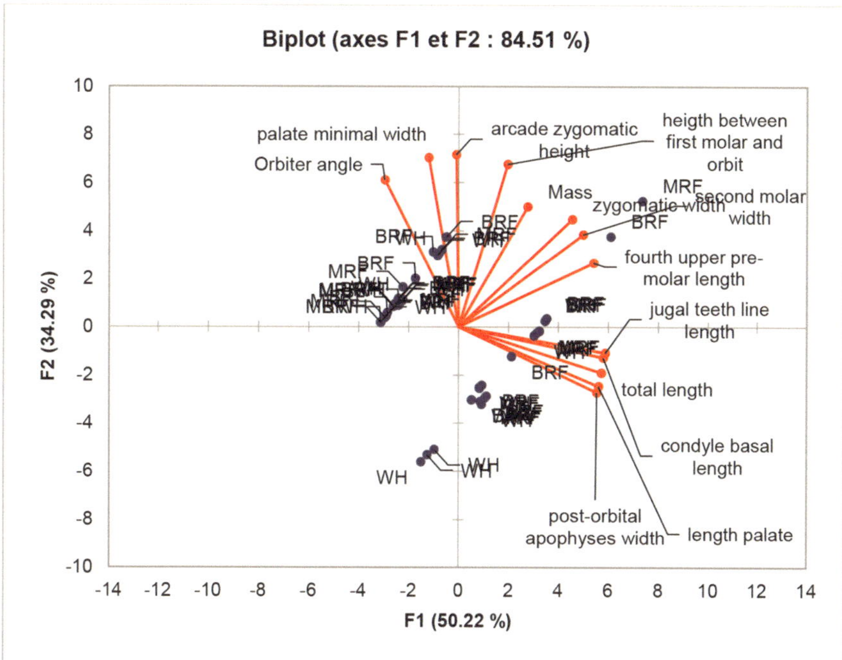

Figure 2.
Grouping of similar cranial measurements by agro-ecological zone.

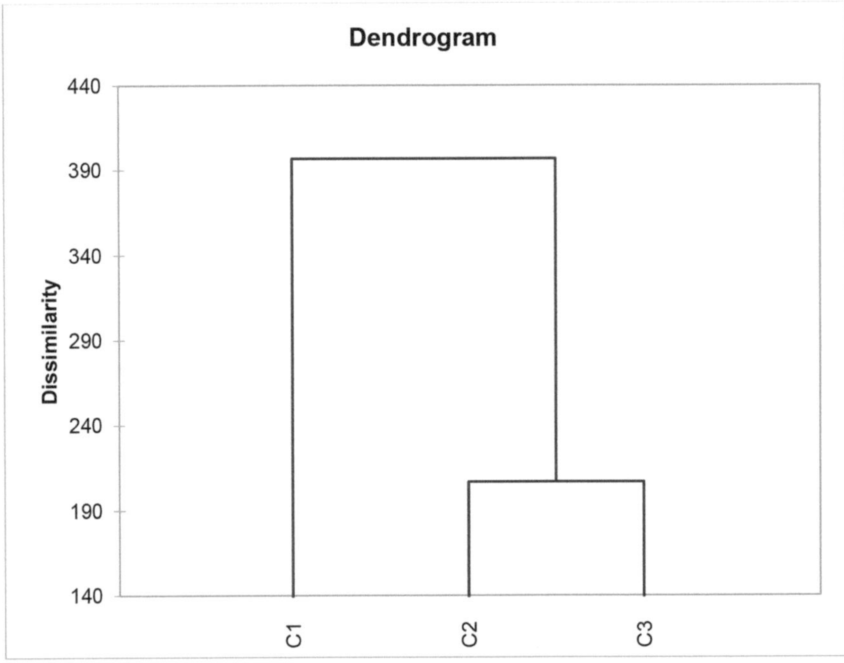

Figure 3.
Dendrogram of cranial measurements of the blue duiker according to cranial measurements. c: subclass of cranial measurements.

3.5 Discriminant factorial analysis (DFA) of blue duiker

The discriminant factor analysis (DFA) of the cranial measurements made it possible to find out which characteristics make it possible to best separate the classes (groups) of individuals by agro-ecological zone. It gives a graphical representation that best accounts for this separation (**Figure 4**).

The explanatory variables that discriminate the three groups are those whose probability is lower than the chosen risk of error (≤5%). Thus, among the characteristics (total length of the skull; zygomatic width; jugal teeth row length; maximum palate width; minimum palatal width; post orbital apophyses width; orbital first molar base height; zygomatic arch height; maximum fourth upper premolar length; maximum width second molar; palatine length; condylo-basal length; pre-maxillary width; orbital angle, mass.) are not influenced by agro-ecological zones. They are grouped into three barycentre's corresponding to the different subclasses which bring together the agro-ecological zones presenting the best similar measurements. However, they can only be discriminated against on the basis of sex. Thus, among the 15 measurement variables, the pairs (total length of the skull; mass), (zygomatic width; pre-maxillary width), (length of cheek teeth row; mass), (length of cheek teeth row; maximum palate width), maximum palate width; orbital angle), (zygomatic arch height; pre-maxillary width), (maximum length of the upper fourth premolar; pre-maxillary width) and (orbital angle; mass) therefore better discriminate between the three subclasses. The sign of the coefficient of the linear discriminant function makes it possible to locate the cranial measurements of each of the subclasses; the negative sign for subclass 1; the positive sign for subclass 2 and the negative sign for subclass 3.

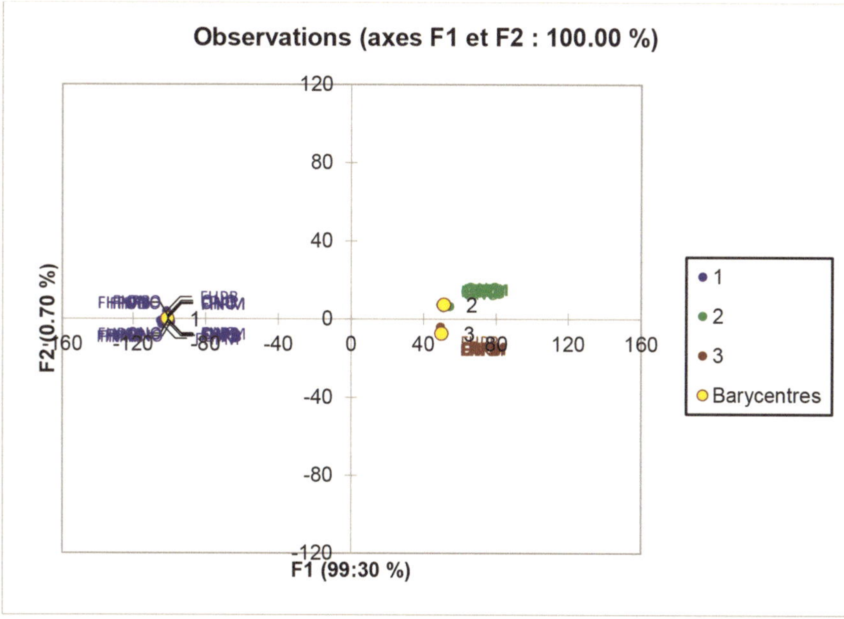

Figure 4.
Discrimination of blue duiker cranial measurements according to agro-ecological zones.

4. Discussions

According to the [29], wildlife species are typically undervalued on the basis of productivity and size compared to domesticated animals. In some contexts, however, wildlife can compete with livestock, particularly given the opportunities they provide, including ecotourism, hunting, meat and other ecosystem benefits. She goes further by asserting that, approaches to managing wildlife, therefore the blue duiker, should include improving knowledge of the use and trade of wild species and an understanding of the ecology of the species concerned. In the same vein, if these conditions were combined and incorporated into sound national wildlife management strategies, it would be possible to achieve more sustainable use of wildlife for food and other purposes. Thus, knowing the population structure of the blue duiker helps to achieve the objectives of the CBD. Biometric characteristics of living and cranial blue duiker show three intra-specific subclasses. This would imply the existence of a blue duiker meta-population common to the three agro-ecological zones. As such, one could suspect three blue duiker subgroups due to the fragmentation of their habitat [30]. The C1, C2 and C3 subclasses would in fact be a blue duiker meta-population but isolated by the reduction and fragmentation of the forest ecosystem. The consequence of this fragmentation is inbreeding which could make the blue duiker population fragile and susceptible to any type of disease [31].

Knowledge of the resource allows good monitoring because in article 17 on the monitoring of the use of genetic resources, the Ngoya Protocol stipulates that, in order to promote compliance with the applicable rules, each Party shall take appropriate measures to monitor the use of genetic resources. Use of genetic resources and increase transparency regarding such use [32]. These measures must be related to the use of genetic resources or the collection of relevant information, among others, at any stage of research, development, innovation, pre-commercialization or commercialization. Sustainable wildlife management refers to the proper management of wildlife species to maintain their populations and habitats over time, taking into account the socio-economic needs of human populations [33]. When sustainably managed, wildlife can provide long-term nutrients and income to indigenous peoples and local communities, contributing significantly to local livelihoods and safeguarding human and environmental health [29, 34, 35].

5. Conclusion

The biometric parameters grouped according to agro-ecological zones are strongly linked to each zone. The characteristics of the living blue duiker gave three groups of characteristics. The same is true for cranial measurements. These three groups of characteristics allow us to say that the blue duiker in Cameroon has three structures corresponding to three subspecies closely linked to their agro-ecological zones. Although similar, these three subspecies have different structures due to their habitat, which varies from one agro-ecological zone to another. The knowledge of these sub-species has an impact on the ecological monitoring of the blue duiker in the sense that the monitoring plan will be built according to each agro-ecological zone thus corresponding to each sub-species.

Each country is called upon to gather information on its genetic resources. This is done at the local level, with the populations living near the forests. To achieve this, it is necessary to develop data collection tools that are accessible to all and easy to use. This data will be used at municipal and national levels to sustainably manage the resource. Biometrics is used to help local populations and resource managers identify the resources in order to easily collect data that can be used to make management decisions.

Acknowledgements

ACPRD (African Center for Participatory Research and Development) fir funding on this research and, hunters, sellers and personal wildlife for the collaborations.

Thanks

African wisdom teaches that it is easier for a man to climb a tree with both arms than with one. This amounts to saying that within the framework of an intellectual work, a single head, even that of a scholar, cannot carry it out alone. This research has benefited from the scientific advice of some, the moral support and the financial support of others. In the impossibility of quoting them all, our thanks go particularly to:

- Pr TSI Evaristus ANGWAFO, Full Professor, at the Faculty of Agronomy and Agricultural Sciences University of Dschang, Department of Forestry, for his contribution to the writing of this article;

- Pr MEUTCHIEYE Félix, Lecturer, teacher at the Faculty of Agronomy and Agricultural Sciences, Department of Zootechnics, for his contribution to the writing of this article;

- The whole FOKAM family for the support they have always given me;

DOI: http://dx.doi.org/10.5772/intechopen.1002909

- The TCHOUASSEP family for their moral support and especially to my wife WOUNGWA TCHOUASSEP Pamela Nichel for her advice and her moral and financial support.

Nomenclature

CBD	convention on biological diversity
COMIFAC	Central African Forests Commission
BRF	biomodal rain forest
FDA	Factorial Discriminant Analysis
HAC	Hierarchy Ascending Classification
MRF	Monomodal Rain Forest
PCA	Principal Component Analysis
WH	Western Highlands
Z	zone

A. Appendix

See **Table A1** and **Figures A1–A3**.

Cranial measurements	r	Constancies and std. error	Coefficients and std. error	Std error of estimation	Mean (mm)
Total length	0.274	114.861 ± 1.540	0.425 ± 0.196	3.854	118.01 ± 3.97
Zygomatic width	0.760	53.145 ± 0.342	0.387 ± 0.043	0.855	56.02 ± 1.30
Length of the row of cheek teeth	0.314	32.855 ± 1.323	0.424 ± 0.168	3.310	36.01 ± 3.45
Maximum palate width	0.120	36.112 ± 4.599	0.537 ± 0.585	11.507	40.10 ± 11.49
Minimum palatal width	0.588	17.672 ± 0.597	0.421 ± 0.76	1.495	20.80 ± 1.83
Width of the post-orbital apophyses	0.191	26.106 ± 2.644	0.497 ± 0.336	6.615	29.80 ± 6.68
Height between the base of the first molar and the orbit	0.756	11.919 ± 0.355	0.397 ± 0.045	0.889	14.87 ± 1.34
Zygomatic arch height	0.599	8.648 ± 0.413	0.300 ± 0.053	1.034	10.87 ± 1.28
Maximum length of the upper fourth premolar	0.592	3.516 ± 0.415	0.295 ± 0.053	1.038	5.71 ± 1.27
Maximum second molar width	0.669	3.030 ± 0.355	0.310 ± 0.045	0.889	5.33 ± 1.18
Palatal length	0.221	57.156 ± 2.304	0.506 ± 0.293	5.764	60.92 ± 5.86
Condylo-basal length	0.316	96.352 ± 1.931	0.623 ± 0.246	4.831	100.98 ± 5.05
Premaxillary width	0.181	13.887 ± 10.375	1.846 ± 1.320	25.956	27.61 ± 26.16
Orbital angle	0.301	26.240 ± 0.826	0.252 ± 0.105	2.066	28.12 ± 2.14

Std: standard.

Table A1.
Model blue duiker cranial identification sheet.

Figure A1.
Stake-fed blue duiker at Campo in Mabiogo village.

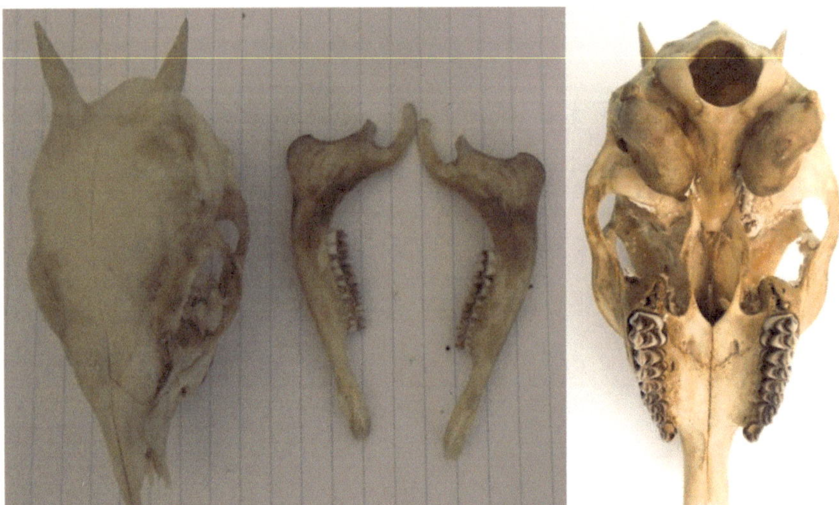

Figure A2.
Male blue duiker skull.

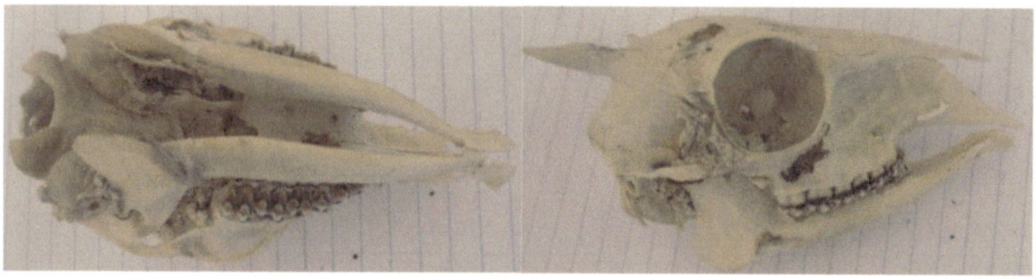

Figure A3.
Female duiker skull.

Author details

Miantsia Fokam Olivier[1,3]*, Felix Meutchieye[1] and Evaristus Tsi Angwafo[2]

1 Biotechnology and Bio-informatics Research Unit, University of Dschang, FASA, Cameroon

2 University of Bamenda, Cameroon

3 University of Bertoua, Higher Institute of Agriculture, Wood, Water Resources, and Environment of Belabo, Cameroon

*Address all correspondence to: miantsiaolivier@gmail.com

References

[1] Sala OE, Chapin FS, Armesto JJ, Berlow E, Bloomfield J, Dirzo R, et al. Global biodiversity scenarios for the year 2100. Science (New York, N.Y.). 2000; **287**(5459):1770-1774

[2] Nazarevich V. The sixth species extinction event by humans. Convergence Earth Common Journal Convergence MacEwan University. 2015;**5**(1):10

[3] Ripple WJ, Abernethy K, Betts MG, Chapron G, Dirzo R, Galetti M, et al. Bushmeat hunting and extinction risk to the world's mammals. Royal Society Open Science. 2016;**3**(10):160-498

[4] Duraiappah AK, Naeem S, Agardy T, Ash NJ, Cooper HD, Díaz S, et al. Millennium ecosystem assessment: Ecosystems and human wellbeing. Ecosystems. 2005:1-100

[5] Hooper DU, Adair EC, Cardinale BJ, Byrnes JEK, Hungate BA, Matulich KL, et al. A global synthesis reveals biodiversity loss as a major driver of ecosystem change. Nature. 2012; **486**(7401):105-108

[6] Bahuchet S. Le rôle de la restauration de rue dans l'approvisionnement des villes en viande sauvage: le cas de Yaoundé (Cameroun). Travaux de la Société d'Ecologie Humaine. 2000:171-182

[7] COMIFAC. Stratégie sous régionale pour l'utilisation durable de la faune sauvage par les communautés autochtones et locales des pays de l'espace COMIFAC. Adoptée en Conseil des Ministres de la COMIFAC en janvier 2015. Rapport; 2015. p. 25

[8] Hette S. Quantification de la viande de brousse prélevée et consommée dans trois villages du sud-est du Cameroun. Travail de fin d'études présenté en vue de l'obtention du diplôme de master bio ingénieur en gestion des forêts et des espaces naturels. Liège université, Agro-Bio-Tech; 2018. p. 75

[9] FAO. Caractérisation phénotypique des ressources génétiques animales. Directives FAO: sur la production et la santé animales; 2013. p. 151

[10] Baumung R, Simianer H, Hoffmann I. Genetic diversity studies in farm animals–A survey. Journal of Animal Breeding and Genetics. 2004; **121**:361-373

[11] Mahammi F. Caractérisation phénotypique et moléculaire des populations de poules locales (Gallus gallus domesticus) de l'Ouest Algérien. Université des Sciences et de la Technologie d'Oran « Mohamed Boudiaf ». Faculté des Sciences de la Nature et de la Vie Département de Génétique Moléculaire Appliquée Thèse présentée En vue de l'obtention du Diplôme de Doctorat; 2015. p. 180

[12] Van Vuuren BJ, Robinson TJ, Retrieval of four adaptive lineages in duiker antelope: Evidence from mitochondrial DNA sequences and fluorescence *in situ* hybridization. Molecular of Phylogenetic Evolution. Elsevier. 2001;**20**:409-425

[13] Bennun L. Davies G. Howell K. Newing H Linkie M: La biodiversité des forêts d'Afrique: Manuel pratique de recensement des vertébrés ; 2004. p. 180

[14] Miantsia FO. Analyse situationnelle et perspectives de game-ranching d'ongulés sauvages (*Cephalophus* spp. et *Potamochoerus porcus*) au Cameroun [Thèse présentée en vue de l'obtention du diplôme de Master Recherche en

gestion de l'environnement, option gestion des ressources naturelles]. Département de Foresterie, Université de Dschang; 2016. p. 87

[15] Dubost G. The size of African forest artiodactyls as determined by the vegetation structure. African Journal of Ecology. 1979;**17**(1):1-17

[16] Dubost G. Comparison of the diets of frugivorous Forest ruminants of Gabon. Journal of Mammalogy. 1984;**65**(2): 298-316

[17] Gauteir-Hion A, Emmons LH, Dubost G. A comparison of the diets of three major groups of primary consumers of Gabon (primates, squirrels and ruminants). Oecologia (Berl.). 1980; **45**:182-189

[18] Nasi R, Van Vliet N. Mesure de l'abondance des populations d'animaux sauvages dans les concessions forestières d'Afrique centrale. Unasylva. 2011; **62**(2):49-55

[19] Towa OW, Bobo KS, Djekda D, Keumbeng B, Bobo R, Moaga Y, et al. Population density estimates of forest duikers (*Philantomba monticola* & *Cephalophus* spp.) differ greatly between survey methods. African Journal of Ecology. 2018;**56**(4):908-916

[20] Vlaeva R, Georgieva S, Barzev G, Ivanova I. Morphological and phenotypic characteristics of donkeys in some regions of Bulgaria. Trakia Journal of Sciences. 2016;**1**:92-95

[21] Dubost G. L'écologie et la vie sociale du Céphalophe bleu (*Cephalophus monticola* Thunberg), petit ruminant forestier africain. Zeitschrift für Tierpsychologie. Ethology. 1980;**54**: 205-266

[22] IRAD. Deuxième rapport sur l'état des ressources phytogénétique pour l'alimentation et l'agriculture au Cameroun; 2008. p. 93

[23] Ducos, Etude d'un squelette de taurin kapsiki: note préliminaireIn: Des taurins et des hommes : Cameroun, Nigéria [en ligne]. Marseille: IRD Editions; 1998. DOI: 10.4000/books. irdeditions.5403

[24] Moulay AB, ZNARI M, Teresa A. Comparative study of the cranial fluctuating asymmetry in two Dorcas gazelle subspecies Gazella dorcas massaesyla vs. G. d. neglecta). Bulletin de l'Institut Scientifique, Rabat, Section Sciences de la Vie. 2018;(40):1-9

[25] Lajoie A. Philibert A et Jolicoeur H : Guide de prises de mesures crâniennes pour des fins de taxonomie et d'identification des canidés. Ministère des Ressources naturelles de la Faune et des Parcs. Direction du développement de la faune. Québec: Bibliothéque nationale du Québec; 2003. p. 34

[26] Lahm SA. Utilization of forest resources and local variation of wildlife populations in northeastern Gabon. In: Hladik CM, Hladik A, Linares OF, Pagezy H, Semple A, Hadley M, editors. Tropical Forests People and Food, MAB Series. Vol. 13. UNESCO; 1993. pp. 213, 852-226

[27] Khaldi Z. Haddad B. Soui S. Rouissi H. Ben Gara A et Rekik B. Caracterisation phenotypique de la population Ovine du Sud Ouest de la Tunisie. Animal Genetic Resources. **49**(0):1-8. DOI: 10.1017/ S2078633611000361

[28] Miantsia FO, Meutchieye F, Tsi AE. Inedible diversity use of blue duiker, Cephalophus Monticola (Thunberg, 1789) of Cameroon. JOJ Wildlife & Biodiversity. Juniper. 2021;**4**(2):555631. DOI: 10.19080/CTBEB.2021.04.555631

[29] MINEPDED. Loi n° 2021/014 du 9 juillet 2021 régissant l'accés aux ressources génétiques, à leurs dérivés, aux connaissances traditionnelles associées et le partage juste et équitable des avantages issus de leur utilisation. p. 19

[30] Van Vliet N, Nasi R. Mise en évidence des facteurs du paysage agissant sur la répartition de la faune dans une concession forestière. Diversité Biologique. Faune Gabon. Bois et Forêts des Tropiques. 2007;**292**(2):23-37

[31] Quagleitti B. Impact de la consanguinité et de l'hybridation chez quatre auxiliaires de lutte biologique. Sciences biologiques [Thèse de doctorat]. Université de Côte d'Azur; 2017. p. 180

[32] CBD. L'accès aux ressources génétiques et le partage juste et équitable des avantages découlant de leur utilisation relative à la Convention sur la diversité biologique. Protocole de Ngoya. 2012:16

[33] Starkey M, Scholtz O, Taylor G. Wildlife monitoring practices and use in Central Africa. In: Program on African Protected Areas and Conservation (PAPACO). Wildlife Conservation Society; 2015. Available from: https://papaco.org/wp-content/uploads/2015/09/IUCN-Monitoring-PA-2-in-Central-Africa.pdf vue le 04/04/2021

[34] FAO. Gestion durable des forêts et de la faune sauvage en Afrique: Améliorer la valeur, les avantages et les services. Nature & Faune. 2016;**30**(2): 107

[35] FAO. Projet FAO/GEF "Gestion durable du secteur de la faune sauvage et de la viande de brousse en Afrique centrale". In: Rapport Final. 2017. p. 16